W9-BEU-197

About Island Press

Island Press is the only nonprofit organization in the United States whose principal purpose is the publication of books on environmental issues and natural resource management. We provide solutions-oriented information to professionals, public officials, business and community leaders, and concerned citizens who are shaping responses to environmental problems.

In 2004, Island Press celebrates its twentieth anniversary as the leading provider of timely and practical books that take a multidisciplinary approach to critical environmental concerns. Our growing list of titles reflects our commitment to bringing the best of an expanding body of literature to the environmental community throughout North America and the world.

Support for Island Press is provided by the Agua Fund, Brainerd Foundation, Geraldine R. Dodge Foundation, Doris Duke Charitable Foundation, Educational Foundation of America, The Ford Foundation, The George Gund Foundation, The William and Flora Hewlett Foundation, Henry Luce Foundation, The John D. and Catherine T. MacArthur Foundation, The Andrew W. Mellon Foundation, The Curtis and Edith Munson Foundation, National Environmental Trust, The New-Land Foundation, Oak Foundation, The Overbrook Foundation, The David and Lucile Packard Foundation, The Pew Charitable Trusts, The Rockefeller Foundation, The Winslow Foundation, and other generous donors.

The opinions expressed in this book are those of the author(s) and do not necessarily reflect the views of these foundations.

About The Environmental Careers Organization (ECO)

The ECO Guide to Careers that Make a Difference is just one way that The Environmental Careers Organization helps college students and job seekers. ECO is a private, national, nonprofit organization committed to the development of a diverse network of professionals and leaders who are creating a more sustainable world. ECO has been serving the nation with internships, fellowships, leadership education, diversity recruiting, conferences, publications, and other services since 1972. To learn more about ECO, or to share your thoughts about this book, visit www.eco.org.

THE ECO GUIDE TO CAREERS THAT MAKE A DIFFERENCE

THE ENVIRONMENTAL CAREERS ORGANIZATION

Beth Ginsberg
Editor and Project Director

Kevin Doyle
National Program Director

John R. Cook Jr.
President

THE ECO GUIDE TO CAREERS

that Make a Difference

ISLAND PRESS

Washington • Covelo • London

Library of Congress Cataloging-in-Publication data.

The ECO guide to careers that make a difference.
    p.   cm.
 Includes bibliographical references and index.
 ISBN 1-55963-967-9 (pbk. : alk. paper) — ISBN 1-55963-966-0
(cloth : alk. paper)
  1. Environmental sciences—Vocational guidance.   I. Environmental
Careers Organization.
  GE60.E34 2004
  628—dc22

                                         2004024493

British Cataloguing-in-Publication data available.

Printed on recycled, acid-free paper

Design by Teresa Bonner

Manufactured in the United States of America

10 9 8 7 6 5 4 3 2 1

Contents

Preface

You cannot save the land apart from the people
or the people apart from the land.
—Wendell Berry

Congratulations! By picking up this book, you have demonstrated a desire to create a career that makes a positive difference in the world. Throughout the nation and across the globe, millions of people are working to develop a new kind of society and economy—a sustainable way of life that promotes ecological health, economic security, and social justice for all people, and for the other species with whom we share the earth. This work is the great project of the twenty-first century, and the talent and commitment of people like you will determine whether or not it will succeed.

Some people are drawn to this mission by a passion for "the land"—the natural world. They are attracted by their love for animals, birds, fish, and plants, or by the power and beauty of the world's forests, plains, mountains, and oceans. Many others find their calling in a deep concern for "the people." They are looking for work that promotes human health and well-being and that will reverse past and current injustices of race, class, ethnicity, gender, and geography. Finding a sustainable path, as Berry reminds us, will require equal attention to both land and people.

Whatever your motivation, welcome! Although the conservation movement in the United States has been with us for more than a century and modern environmentalism is over forty years old, the emerging global network for a more sustainable society is still in its childhood. The next years of your career are linked with a very special moment in human history.

There is much to be done everywhere and on every scale, from local neighborhoods of a few square blocks to immense natural areas measured in millions of acres. We must preserve and protect our air, land, and waters; restore damaged ecosystems; redesign our towns and cities; decrease the ever-widening gap between rich and poor; and find creative solutions to dozens of other interlocking problems.

With so much to be done, there is literally something for everyone. Scientists and engineers are desperately needed, both from traditional fields and new, interdisciplinary programs. There are opportunities for economists, lawyers, educators, entrepreneurs, activists, managers, technicians, computer specialists, and architects. Career opportunities are abundant in government, business, educational institutions, and the nonprofit community. We can also predict with certainty that the next two or three decades will call for professions that haven't been invented—or even dreamed of—today.

Departure from *The Complete Guide to Environmental Careers*

The Environmental Careers Organization is well known as the author of *The Complete Guide to Environmental Careers*, a single-volume reference book also published by Island Press. Published for the first time in 1989 and completely updated in 1993 and 1998, *The Complete Guide to Environmental Careers* has been a valuable resource for students, teachers, job seekers, and career advisors.

Those who love that book and are eager for a new edition should be aware that *The ECO Guide to Careers that Make a Difference* is a very different title and has different aims. Whereas the previous ECO books were reference books aimed at providing a comprehensive review of environmental careers, statistics, and job-hunting strategies, the book you are holding now seeks to focus intensively on the issues facing our world and the strategies and tactics environmental professionals are using to find creative solutions. We have called upon some of the nation's best-known leaders, thinkers, and managers to help us. Although readers will find the valuable career information they need to get started, the world of environmental work has become so large that a single reference volume is no longer possible. Instead, this guide is intended to introduce you to the many different possibilities that await you. More detailed information on careers in the government, nonprofit agencies, and the private sector will be provided in subsequent volumes.

How to Use This Book

The ECO Guide to Careers that Make a Difference is unlike any other environmental careers book. Rather than tromping through a list of job titles or professional disciplines, *The ECO Guide* is based on the premise that people like you are motivated by a deep concern about the critical issues facing our planet and our people. You do not want to choose between making a difference and making a living.

To offer practical help, we have identified nineteen of our most intractable environmental, conservation, and sustainable development concerns—issues like climate change, water quality, forestry, wasteful consumption, biodiversity loss, energy production, and urban sprawl. These are the problems that dominate national headlines, legislative sessions, and academic research agendas—and these are the problems that inspire people to ask, "How can I put my career to work towards a positive solution?"

• ## Conversations with the Experts

For each of these nineteen concerns, we sought out a recognized expert with a deep involvement in the issue and a detailed understanding of the strategies and tactics being employed by career professionals. We engaged each of these experts in a lengthy conversation designed to answer four questions:

1. What is the nature of the problem?
2. How are environmental professionals approaching this problem? What kind of strategies and tactics are they using?
3. What advice do you have for people who want to get involved?
4. Are there resources people should know about to learn more?

The ECO Guide team then collaborated with each of the experts to arrive at the edited "conversations" in this book. In some cases, our experts took charge and wrote the text you're about to read. In others, they shared ideas back and forth with editor Beth Ginsberg. Some encouraged us to add significant ideas of our own. It was truly a conversation, and we are deeply thankful for the time and thoughtfulness given to us by these extraordinarily busy people.

- Featured Jobs

To bring *The ECO Guide* even further down-to-earth, we surveyed leaders across the country to identify forty environmental and sustainability jobs that capture the diverse spirit of this work and offer a snapshot of what people do, how much they are paid, and how they got there. (Featured jobs are indicated in **boldface type** throughout the text.) Our forty selections are not intended as a comprehensive list. The world of environmental work is vast and diverse, and for every job title we've selected, there are many more. Our hope is to be representative of the kinds of careers that await you. For more information about environmental, conservation, natural resource management, and sustainable economy job titles mentioned in this book, visit the ECO website at www.eco.org.

- Career Spotlights

Finally, using ECO's own unique networks in the environmental world, we have identified particularly effective businesses, government agencies, nonprofit groups, and academic institutions to serve as "Career Spotlights." These spotlights help bring to life the organizations and professionals who are doing good work for the planet. We hope that you will find them representative of many similar employers throughout the nation. The city of Santa Monica, California, profiled in chapter 16, for example, is certainly doing leadership work in building a sustainable community. And yet, we think that Dean Kubani, the city's senior environmental analyst and sustainable city coordinator, will be the first to agree when we say that there are great career opportunities in local government right in your own community.

The ECO Guide can be used in a number of ways. It's a brief yet information-packed introduction to environmental work for people from a wide variety of disciplines who want to launch a meaningful career and make an impact on these vexing issues. You might use *The ECO Guide* as a reference book for finding answers to questions about specific issues and opportunities. You might read certain chapters as entry to additional sources of information and insight. Alternatively, you might choose to read it straight through to gain a comprehensive introduction to key environmental issues, the ways they are linked to one another in unexpected and unpredictable ways, and how to develop a career of your own that addresses them.

We encourage you to make *The ECO Guide* your own book and to use it as a source of information, inspiration, and involvement with a creative network of leading professionals, academics, activists, and leaders. All over the world, this remarkable network of women and men is hard at work on the issues described in *The ECO Guide to Careers that Make a Difference*. We're glad that you've decided to link your own career to theirs. We wish you all the best in the coming years. And, on behalf of both the land and the people, thank you!

John R. Cook Jr.
Founder and President
The Environmental Careers Organization
Boston, Massachusetts

Acknowledgments

Like the environmental work it chronicles, *The ECO Guide to Careers that Make a Difference* required the labors of many different people. Renowned experts shared their time, ideas, and stories with us, and their often-unheralded staff people were extraordinarily helpful and supportive. Colleagues and friends at professional associations, government agencies, businesses, and nonprofit groups provided insights that were unavailable anywhere else. Dozens of working professionals helped us understand the realities of twenty-first-century environmental problem-solving so that we could convey realistic and accurate information to you—a new generation of leaders and professionals. Only a few of those who contributed can be mentioned by name here, but our thanks go out to every person who helped us turn the idea of this book into a reality.

The ECO Guide to Careers that Make a Difference is the most recent evolution of *The Complete Guide to Environmental Careers* (Island Press 1989, 1993, 1998). Although not meant to fill the same niche as that volume, we owe a large debt to the teams that created previous ECO books. The first edition of *The Complete Guide* was managed by the Environmental Careers Organization's former Great Lakes director, Lee DeAngelis, and co-written by Lee with Stephen Basler. The 1993 edition was managed by Kevin Doyle and co-written by him and Bill Sharp, with assistance from Elaine Appleton, Catherine Pederson, Jean Anderson, and Lee DeAngelis. The 1998 edition was managed, edited, and partially written by Kevin Doyle, with assistance from Tanya Stubbs, Sam Heizmann, Dylan Murray, and Meryl Brott. Many thanks also go to the George Gund Foundation and the Pew Charitable Trusts for the financial support that made the first edition of *The Complete Guide to Environmental Careers* possible.

The Complete Guide has become the nation's best-selling book about environmental jobs, and *The ECO Guide to Careers that Make a Difference* would not exist today without the success of these previous teams. We hope this book meets the standards they set and accurately reflects evolving opportunities for environmental careers.

The ECO Guide to Careers that Make a Difference is a new and exciting book—a departure from our previous career publications. This project was directed by ECO executive Kevin Doyle and edited and coordinated by Beth Ginsberg. Nicole Sequino provided great support to the featured jobs that appear in every chapter. Alicia Dunn's contributions to the featured jobs and her support for the entire initiative made her an important member of the team. We would also like to thank Dhara Vala for her help with the Career Spotlights.

Although creating this book was a team project (as these acknowledgments make clear), the heart and soul of *The ECO Guide* is Beth Ginsberg. This is primarily her book, and it reflects her passionate commitment to the sustainable communities movement around the world and especially to those who labor at the local level. Beth came to us straight out of school, and for nearly two years as ECO's staff writer she guided and sustained the project, prodded and cajoled our contributors, improved the prose of others, and wrote compelling text of her own. Her unique style, personality, and intelligence are on each page of *The ECO Guide*, and her professionalism and good humor have been sparks to everyone in the ECO community. Thank you, Beth!

We would like to extend our deepest gratitude to the twenty experts who participated in conversations on which the chapters are based and whose continued efforts in the writing and editing processes were invaluable: Robert Bullard, Tom Cackette, Don Chen, Eileen Claussen, Theo Colborn, Kevin Coyle, Seth Dunn, Jerry Franklin, Michel Gelobter, Stuart Hart, Martha Honey, Fred Kirschenmann, Whitney Montague, William McDonough, Stuart Pimm, Sandra Postel, Ellen Prager, Carl Safina, Betsy Taylor, and Patricia-Anne Tom. Thank you to the many people on each of these experts' staff who provided assistance throughout the project, especially Pam Clymer, Amy Rutledge, and Sophie Mintier.

We would like to further thank Eileen Claussen and Sarah Cottrell for going beyond the requested participation for the climate change chapter and Theo Colborn for her time-intensive work. Thank you to Fred Kirschenmann and Patricia-Anne Tom for their timely contributions. Special thanks to Chris Reiter for his much appreciated help with the architecture chapter editing, as well as to Kevin Coyle for helping finish the education chapter.

We would also like to thank the many other environmental professionals featured throughout the book. Without their participation, we would be missing the perspectives of those who work in the field day in and day out, doing the very work that our readers are most interested in and most likely to seek out. We applaud and continually support your efforts in making daily differences.

Special thanks to Todd Baldwin at Island Press, who is everything one could want in an editor, and to James Nuzum for his dedicated assistance throughout the entire project. Both of their efforts in managing this project and helping guide its conceptualization have been extraordinary. *The ECO Guide* would not be what it is today without them! Additional thanks goes to our production editor, Cecilia González.

The ECO Guide to Careers that Make a Difference is a project of the Environmental Careers Organization and draws on the strength of the people who manage ECO's respected environmental intern program, national career conference, and other programs. Thank you to the 2002–2004 staff in Boston, San Francisco, and Seattle for their dedication to environmental career education. Your work is a source of daily inspiration to me.

Finally, I would like to dedicate this book to the more than 9,000 men and women who have begun environmental careers as ECO interns over the last thirty years and to the generous and thoughtful environmental professionals who serve as project advisors and mentors. I started the Environmental Careers Organization in 1972 with a vision that college students and recent graduates could make a difference in their communities if they worked together with today's leaders on well-designed projects. More than thirty years and thousands of interns later, that vision is alive and well. My thanks to everyone who has helped make my internship the best one of all.

John R. Cook Jr.
Founder and President
The Environmental Careers Organization
Boston, Massachusetts

Featured Jobs

Career Spotlights

Introduction

Inventing the Twenty-First-Century Sustainable Economy

As an aspiring environmental professional, by now you've probably heard many calls for "sustainable development" around the world. And those calls will get only more urgent. The next fifty years represent a critical period in our efforts to protect what remains of wild nature, restore ecological health to places that are critically out of balance, and reverse long-standing trends that threaten both people and wildlife.

If, however, you have a hard time imagining exactly how your career might unfold in "the twenty-first-century sustainable economy"—or even what such an economy might look like—don't worry. You're not alone. The concept of sustainability is hotly debated and hard to define, even by its most ardent supporters. One study found over sixty different definitions of the term. And conflicting visions of a sustainable future come attached to starkly different recommendations for policies and institutional changes that might get us there. Scholars and advocates may speak in confident agreement about a future economy measured by the "triple bottom line"—where environmental, economic, and social issues are integrated and given equal weight—but real progress on the ground is much less clear, especially in the United States.

Even so, the vision of a sustainable economy is rapidly being accepted as the primary goal of environmental work and as the conceptual framework within which scientists, engineers, businesspeople, policy experts, educators, and others will work together. With this in mind, an essential first step for every aspiring environmental professional is to develop your own conception of sustainability—what it is and what it means for your professional and career choices.

One could do worse than to begin with the definition from *Our Common Future*, an influential 1987 report from the United Nations that defined sustain-

1

able development as "development that meets the needs of the present without compromising the ability of future generations to meet their own needs."

In creating a twenty-first-century economy that will bring this definition into the forefront of our society, what might such development look like in a real community? Jeff Hollender, CEO of Seventh Generation, the nation's leading marketer of natural household products, urges us to take a look at an experiment called EcoPark in Burlington, Vermont, in a chapter contributed to the 2002 book *Sustainable Planet*. EcoPark is a joint project of the Intervale Foundation and the city of Burlington. Located on an 800-acre floodplain along the banks of the Winooski River, EcoPark seeks to emulate what Hollender calls a "natural paradise," in which the river "drains distant mountain rains into Lake Champlain and cattail, heron, oak, and blackbirds coexist within a marsh, forest, river and field." It is modeled on the self-regulating ecosystems that supported and enhanced the planet long before human interference began to degrade their capacities; its designers call it "agriculture-based industrial ecology."

But how does it work? By definition, an eco-park is a cluster of profitable businesses that collaborate on waste, energy, water, and materials management to minimize their environmental impact. According to Hollender, the EcoPark experiment exemplifies such ideas, as the wastes of one business on the lot are fed into another. Among the project's most important innovations is a "Living Machine," in which a "labyrinth of enormously connected water tanks, each containing a marine ecosystem, transforms wastes normally discharged into the air, soil, and water, into useful things." For example, a local brewer uses grains grown in the fields. The brewer's by-products become raw materials for a local baker. The baker's wastes become fuel for the "Living Machine," which cleans EcoPark's wastewater and produces fish for the city of Burlington.

This experiment and our vision for a sustainable society are very different, of course, from the current world in which the pursuit of economic growth and material goods dominates both individual and institutional actions and accomplishments, and that also leads to the degradation of both the natural environment and the social services and infrastructures that ensure people's basic needs are met.

In the sustainable world we seek to create, people and the continual effort to improve their physical, social, environmental, and economic well-being are at the center of development efforts. Well-being is the measurement of global progress. Natural resources are protected and valued. Infrastructure design prioritizes efficient energy and water use, and reduces waste. Mixed-use, compact development brings people together and preserves open space. Strong local economies guide human activities and decision making. Looking at the larger picture, the imagined world is one in which we collectively manage local and global resources within ecological limits and also where we enable citizens to meet their economic needs while promoting equity and justice.

This is a world that rests on the tenets of sustainability. And it is not as far away as you might think. The modern environmental movement that began in the 1960s and 1970s is now changing to reflect a more integrated and holistic approach to problem solving. In fact, the dream of a sustainable global society and economy is causing profound changes in how we think about environmental issues and environmental work.

Environmental professionals are now thinking about questions that might at first seem unrelated to "environmental" concerns. Questions like: At what cost is environmental devastation occurring? What is the link between environmental degradation and national security? Do more material assets mean more happiness? How does the gap between the wealthy and the poverty-stricken affect larger global issues? In answering such questions, we realize that, as Lester Brown, president of the Earth Policy Institute, says in *Eco-Economy*, "If we want economic progress to continue, we have little choice but to systematically restructure the global economy in order to make it environmentally sustainable."

This type of thinking embodies the starter seeds for growing an environmental movement that values equity, justice, empowerment, and traditional environmental protection, preservation, and conservation. Environmental professionals will help accomplish the changes necessary to achieve a newly envisioned world.

By definition, building a sustainable world involves the participation of all people—citizens and professionals. For example, when Natural Resources Defense Council (NRDC) senior scientist Allen Hershkowitz proposed a world-scale recycled-paper mill in the South Bronx, he employed many different professionals and community members to create the infrastructure and innovation necessary to create the plant itself and its recycled outputs. He knew that without land use planners, his engineers could not plan sewer infrastructure necessary for reclaiming the sewage treatment plant wastewater to improve the area watershed quality. Without environmental regulators, water resource managers would not be able to ensure the water quality in the newly developed brownfield. Without responsible business people and entrepreneurs to market and sell environmentally sound products effectively, it wouldn't matter that industrial ecologists could design, build, and operate a facility that would help

3

heal a scarred part of the planet. And without citizen participation, it could not be responsive to community needs.

It's informative that Hershkowitz's dream hasn't come to fruition. Often, many more businesses fail than succeed in our current unsustainable society. We can expect that the same will be true in one based on ecological principles. However, the central theme of Vermont's "work-in-progress" EcoPark, Lester Brown's research, and Allen Hershkowitz's experiment in the Bronx is that environmental work in the twenty-first century is *inextricably* linked with developing a sustainable economy.

Today, there are thousands of environmental professionals pursuing this dream. They are learning from what works and what doesn't. Instead of working to control and regulate as they did in the past, they're now working to preserve our natural capital as well as collectively rethink and reinvent the way we live. They are working to create dramatic improvements in technology, protect plant and animal habitats, conserve energy sources, develop renewables, establish market incentives that complement government action, construct better education and ecological understanding, and form greater social and racial justice. In this regard, every area of work—agriculture, transportation, manufacturing, energy, services, housing, construction, health care, mining, fisheries, forestry, water management, finance, education, urban design—has an environmental component.

Environmental work is also no longer thought of as an "add-on," to be tackled separately from other factors. For example, fisheries depletion is no longer a separate problem from climate change, as we have seen ocean warming threaten the habitats of cold-water fish like trout and salmon. Similarly, poor air quality cannot be discussed without some relation to transportation and land use patterns. Norco, Louisiana, is a good example; here, parks are located across the street from big refineries where diesel vehicles are constantly trucking materials back and forth, significantly contributing to child respiratory problems in the area.

There are other connections. Biodiversity loss can be traced directly back to agricultural chemical use, as we have seen in the Dead Zone in the Gulf of Mexico, an area where excess runoff of nutrients such as nitrogen, phosphorus, and fertilizers cause algal blooms that use up all the oxygen in the water. Our mounting waste problems must be considered a reflection of the way we use the very resources we should be protecting—like virgin forests in the Pacific Northwest that we are denuding rather than protecting, only to have them be turned into paper that is thrown away rather than recycled; or depleting nonrenewable energy sources like coal rather than devoting resources to developing alternatives. There are many, many more examples—and you'll find them throughout this book.

Before exploring in detail how environmental professionals can help guide us to a more sustainable world, a little history is in order. How did we get here?

Five Waves of the Environmental Movement

Our current focus on creating a sustainable global economy is the latest "wave" for an evolving environmental movement in the United States that is over 150 years old. Historians of environmentalism have used many different labels and dates to characterize the movement's past waves, but the periods identified by Sherburne Abbott, chief international officer at the American Institute for the Advancement of Science, are particularly well-suited to understanding how environmental careers have evolved along with the political and cultural changes that have fueled environmental concern in this country.

The first wave, which began in the 1850s and peaked in the 1890s, was inspired by a romantic notion of wild nature and the role that wilderness played in forming our nation's character. As wilderness disappeared, poets, painters, politicians, and prophets pushed for the protection of the natural wonders they loved—and feared might be lost. In 1872, Yellowstone National Park in Wyoming became the nation's first national park, and other designations were soon to follow. Venerable organizations like the Sierra Club were born, and Congress set aside the refuges that would later become part of the U.S. Fish and Wildlife Service.

If concern about protecting areas of spectacular scenic beauty formed the basis of the first environmental era, the second wave focused on less spiritual worries. It has been labeled the era of "natural resource management" because of its focus on unbridled misuse of nature's economic bounty. From the 1890s until as late as the 1950s, ever-greater attention was paid to the need for long-term thinking about the use of resources like water, forests, agricultural land, and wildlife habitat. Many of the scientific and expert management professions that are labeled "environmental" careers today emerged and institutionalized themselves as separate academic fields and professions during this period—often accompanied by the creation or expansion of federal government action. The U.S. Geological Survey, for example, received resources to deliver on the mission Congress imagined in 1879 when it charged the new agency with "serving the nation by providing reliable scientific information to describe and understand the earth, minimize loss of life and property, manage water, biological, energy, and mineral resources, and enhance and protect our quality of life."

Progress was fast and far-reaching. In 1905, the U.S. Forest Service was established within the Department of Agriculture to manage forest reserves. The 1911 Weeks Act appropriated nine million dollars to purchase six million acres of land in the eastern United States; and in 1916, the National Park Organic Act officially created the National Park Service to "conserve scenery, wildlife, and historic objects for future generations."

The Great Depression and the "dust bowl" of the 1930s spurred investment in the growing environmental infrastructure. In 1933, President Franklin Roosevelt established the Civilian Conservation Corps (CCC) to reduce unemployment and preserve the nation's natural resources. Foresters and laymen were among the first of two million professionals who began to work on forestry, flood control, soil erosion, and beautification projects through Roosevelt's attempt to boost the economy and address the needs of the land. Expansion of environmental activity continued immediately after the war with the creation of the Bureau of Land Management in 1946 and increases in the budgets and employee ranks of natural resource management agencies and companies.

Abbott calls the third wave of environmentalism the "ecological movement" —the period between the 1950s and 1970. During this two-decade span, increased scientific and ecological understanding transformed our hopes for and approaches to environmental work. Beyond protecting scenic vistas and using nature's bounty more wisely, many events demonstrated that human activities were threatening both environmental and human health in ways that could no longer be ignored. In 1952, smog was blamed for 4,000 deaths in London, leading to calls for air quality analysis. The year 1955 saw the establishment of a link between asbestos and lung cancer, further connecting environmental issues to health problems. In 1962, Rachel Carson—biologist, writer, ecologist, and advocate for nature—put out a clarion call by writing *Silent Spring*, the book that first exposed the dangers of pesticides to the general public. Carson's book, and the subsequent public outcry, brought words like "ecology" to millions of citizens for the first time. This directed the environmental agenda into areas that could not be addressed by relatively nonthreatening actions like creating a park or encouraging windbreaks on farms to slow erosion. Environmentalism now pointed directly to the unintended side effects of our way of life and the daily activities of people in cities and towns.

The legislative activity that would later explode in the soon-to-arrive fourth wave of environmentalism began to appear. In 1963, Congress passed the first Clean Air Act, allocating $95 million to local, state, and national air pollution control efforts. In 1965, the Water Quality Act arrived, giving federal government power to set water standards in absence of state action. The Solid Waste Disposal Act, the first major solid waste legislation, was passed in the same year. The third wave also brought with it new support for the gains of the previous two environmental eras. In 1964, for example, the Wilderness Act designated 9.1 million acres as wilderness in Arizona, California, Colorado, Idaho, Minnesota, Montana, New Hampshire, New Mexico, Nevada, North Carolina, Oregon, Washington, and Wyoming.

The fourth wave in the history of environmentalism has been called the regulatory movement. Begun in 1970, it blazed across the nation for twenty-five remarkable years and is still largely with us today. This phase brought with it a veritable alphabet soup of legislation and regulation aimed at pollution control and, to a lesser extent, pollution prevention. In 1970, Congress established the Environmental Protection Agency (EPA), and the Clean Air Act was signed into law. The National Environmental Policy Act (NEPA) was also signed during this time, requiring all federal actions to have environmental impact analyses. This law was widely copied at the state level, and the nation was introduced to the need for "environmental impact statements."

A host of other important environmental legislation was signed into law by a series of exceptionally concerned classes of Congress. In a short time, we were given the Clean Water Act, Endangered Species Act, Toxic Substances Control Act, National Forest Management Act, Resource Conservation and Recovery Act, "Superfund," and Federal Lands Policy and Management Act. Wilderness bills were also passed in almost every state throughout these decades, designating thousands of acres as protected land throughout the country.

As the environmental movement passed through the first four waves of action and legislation, professionals consistently worked to bring new ideas and priorities to the institutional creations of the past, adding scientific rigor and the lessons learned from experience to improve results. In 1997, for example, Congress passed the National Wildlife Refuge Improvement Act and set wildlife conservation as the top priority in 500 national refuges to protect these areas against military exercises, jet skiing, livestock grazing, and other activities. The new law provided for more science-based management and long-term planning.

As the twentieth century ended, environmental and conservation professionals could look back on more than a century of truly remarkable progress. Some scholars and observers have called the American environmental movement the most successful social movement in history—and for good reason. In the United States today, it is almost impossible to engage in activities that clearly destroy the natural world without running into a well-constructed web of laws and regulations. In 1999, over two million people were employed in jobs related to conservation, natural resource management, and environmental protection. And yet, in spite of this progress, the United States still leads the world in per capita resource consumption and waste generation.

While we have made stunning progress, the methods and tactics of the first four waves are not sufficient for the new, complex challenge of realizing our future goal—the creation of a sustainable economy. The fifth wave of environmentalism may be said to have begun in 1987, with the publication of *Our Common Future*, which first brought the idea of sustainable development into broad public discourse. An often abused and notoriously difficult to define term (as we have seen), "sustainability" has still become the best concept around which twenty-first-century environmental action can organize itself and measure its success.

The sustainability movement is an evolved form of environmentalism that seeks solutions that integrate ecological health, social justice, and economic security on a variety of scales, from local to global levels. In many ways, the fifth wave of environmentalism represents an understanding that we have gone just about as far as we can in solving "environmental" problems by focusing on them as separate from our other human concerns.

Each one of the four previous "waves" has left behind laws, technical approaches, institutions, professions, and career paths that were appropriate to its mission and goals. None of the new waves fully replaced the work of the generation that preceded it. Instead, each added to what had been in place before. The result is that we now have an awkwardly constructed world of environmental employers and professions, taking approaches that are often as likely to conflict with as they are to reinforce one another.

As we build this next phase of environmentalism, the idea of a sustainable economy seeks to bring cohesion and purpose to our work. The challenge is daunting. There is no clear method for combining traditional "environmental" goals with social justice, economic security, poverty alleviation, and sustainable consumption. We are also aware that we must pursue new, more integrated goals without threatening the specifically environmental gains of the past.

Experienced conservationists like Duke University's John Terbough have already challenged the very possibility of "sustainable development" and raised serious charges that we might sacrifice nature to an unproven idea, while failing to speak up for ideas that work—like the expansion of wildlife parks and refuges that exclude people. However, it's hard to imagine another path. The fifth wave of the environmental movement has emerged from harsh realities in both the developed and the developing world. As the human population inches toward nine billion people by the middle of this century, the old ways simply won't do the job—for people or for wildlife. Although there is no guarantee of success, the ideas, practices, laws, and technologies that are aligned with the pursuit of sustainability seem like our best bet to achieve a healthy global environment.

The Changing Role of Business and Government

As environmentalism has shifted from an external control on the mainstream economy to a transforming force that aims to create a sustainable way of life, a fundamental shift is occurring in our understanding of the traditional roles for both business and government. Business, previously thought of as an almost inevitable adversary to environmental protection—because this sector has been most visibly responsible for cutting, mining, drilling, plowing, building, and polluting—is now among our best hopes for creating change.

Whether this newfound role is embraced with ideological fervor and high hopes or whether it is grudgingly accepted with deep skepticism and low expectations, most leading-edge companies agree that our long-term future depends upon transforming the business world and harnessing its creative power towards establishing a sustainable future. In fact, the most successful and innovative companies today are those that are pursuing sustainability and innovation in their own right, rather than focusing on preventative measures, efficiency improvements, or cleanup activities after impacts or damages have occurred. Additionally, the part of the government character that must work with these businesses is becoming more conflicted—less like a policeman and more like a creative partner.

We see this playing out vividly in the business sector where companies are implementing voluntary greenhouse gas (GHG) emissions reductions. GHG emissions have been proven to be the major contributor to climate change. Businesses are in a unique position to reduce such harms because industry operations are a major source. They are also in a unique position to affect environmental change because they are the only actors in the global economy that have the technology, management, innovation, and financial resources necessary for profiting from such opportunities. The Business Environmental Leadership Council (BELC) is a group of approximately forty leading-edge companies that are demonstrating leadership and a sustainability vision by voluntarily committing to emissions reduction targets.

The Pew Center for Global Climate Change, which coordinates this initiative, says that companies involved in the BELC have made substantial greenhouse gas emissions reductions through a wide range of activities. For example, SC Johnson has pledged to reduce GHG emissions by 5 percent per year through 2005. Toyota Motor Corporation has committed by 2005 to reduce its CO_2 emissions by 5 percent from 1990 levels, and by 2010 it will reduce its emissions by 10 percent from 1990 levels. DuPont has already achieved a 65 percent reduction below 1990 levels—well ahead of its 2010 target—and has pledged to hold its emissions flat in the future. It's a hopeful sign that companies working on voluntary emissions reductions have also improved their operations, reduced energy and production costs, and increased their market share.

Environmentally sound practices are good business largely because a significant part of the general public demands it. Consumer power is putting pressure on multinational corporations to become more accountable for their environmental, social, and economic actions both at home and offshore. The environmental movement would be wise to take advantage of this power. After all, it is consumers who create profitable markets for organic foods, fair trade coffee, and union-made apparel, among other consumer goods that are produced under environmentally and socially just conditions. And these markets are growing rapidly.

Businesses that voluntarily find their way to environmentally sound and socially just products, services, and practices help create an environment in which government can responsibly change its approach to environmental protection. Strict legal and regulatory environmental requirements have been developed because of public demand, low trust in business, and a desire to work to achieve measurable results. In the case of climate change, businesses can expect rigorous regulation, mandated technologies, and stiff enforcement if they fail to make low-carbon emissions a hallmark of good business.

Government and business can find more creative ways to achieve outstanding results—government by setting the bar much higher and using its purchasing power more creatively and business by responding with innovative technologies and products. In the world of climate change, the city of Santa Monica, California, provides an interesting example. The local government is attempting to quantify all GHG emissions and enact aggressive emissions reductions plans. In addition to conducting a GHG emissions analysis, it became the first municipality in the nation to purchase 100 percent of its electricity from renewable sources in 1999, and it also reduced citywide GHG emissions by 5.2 percent between 1990 and 1997. There are many other cities all over the world that have jumped into this policy work, and several states have also developed innovative strategies to meet GHG reduction targets in collaboration with businesses and advocates.

Applauding these actions does not negate the need for strong national and international legal action on climate change, biodiversity, or any of the other pressing environmental issues facing the global community and outlined in this book. What's new is the fact that progressive governments and businesses see economic opportunities in the pursuit of sustainability and are willing—even eager—to create a policy environment that supports the development of a sustainable economy.

Living in Two Worlds

Although it's clear that it is not sufficient to address major ecological issues and twenty-first-century challenges through separate environmental protection agencies, corporate environmental, health, and safety departments, or activist groups exclusively devoted to conservation work, it's equally clear that the "sustainable economy" remains a somewhat distant dream—being born in painful fits and starts. There is still great need to focus on traditional efforts like preserving ecosystems worldwide in order to prevent repeated environmental destruction and exploitation. People entering the workplace now must simultaneously be able to secure jobs and build careers within the existing environmental careers infrastructure, as well as address issues in new, more integrated ways that are informed by the triple bottom line of environmental, economic, and social responsibility. New professionals must imagine careers that promote, for example, renewable energy, while at the same time preparing for the jobs that will emerge when that promotion is successful.

Architect William McDonough points out that this is not an easy task. In the architecture and design field, he says, young professionals know that buildings should be designed in harmony with their surrounding environments. They are eager to implement innovative ideas in the real world to accomplish this goal, often going beyond by suggesting ways in which buildings might *enhance* the natural environment. However, unless new ideas have been proven, senior managers with more traditional backgrounds and experience are not likely to let younger workers try out new design concepts, with the exception of risk-taking firms, which are very few. This often puts the idealistic architect in a position where he or she is unable to put past training to use, and structures may end up unsustainable in the long term—a disservice to professionals as well as the environment and our greater society.

Even professionals in recently transformed environmental industries are buffeted by the pressures of further progress toward sustainability. People in the waste management and recycling field, for example, are already thought of as those who collect, sort, and process solid waste and recyclable goods. They also remanufacture and market new products from recycled materials, which is traditionally the best-case end use for a product. However, this end use is a narrow perception of how we can continually reuse products. Instead of downcycling metals or other commonly recycled materials, innovative waste management professionals are calling for the complete redesign and reuse of materials and products so that they fall into two disposal categories: what McDonough and chemist Michael Braungart call "biological nutrients"—materials and products that can be returned safely to the earth in efforts as simple as composting; and "technical nutrients"—materials that can be continually used in closed-loop systems informed by the ideas of industrial ecology. Unfortunately, even mechanical engineers who redesign products and recycling coordinators who work on alternative waste diversion tactics are not yet able to put these ideas into practice. The supporting infrastructure and public will to actually change disposal options and citizen use are not yet in place. Other fields, from biodiversity protection to climate change, are similarly between two worlds.

Conclusion

In looking at the world of environmental careers today, we hope this book outlines the current challenges new professionals will face as our society moves towards a sustainable world. We can safely say there is enough work ahead to keep environmental professionals busy for a lifetime. By understanding the ideas outlined in the chapters of this book (as well as your place in working on those issues), you will not only be on a faster track to obtaining a job, but also be working to make a difference. And by working on multiple levels toward a common goal, hopefully everyone's contributions will one day add up to a sustainable world. Welcome to the new world of environmental careers!

Sector by Sector Snapshot of Environmental Careers Today

What is the current status of "green" employment in the United States? As we've seen, the movement for a more sustainable economy has begun to reach beyond traditional conservation, environmental protection, and natural resource management institutions. The basic ideas of sustainability, however, are still far from being fully incorporated into our mainstream society. With this in mind, where are the jobs today for people who want to combine ecological and social goals? The overview below provides a snapshot of current trends at some leading public and private environmental employers, and suggests directions for job seekers and career planners.

Public Sector

- Federal Government

From the 1970s through the early 1990s, the federal government was clearly a primary source of national environmental progress. Since that time, environmental concerns seem to have slipped from the top of Washington's priority list. Congressional leaders appear to be skeptical about (if not hostile to) the expansion of environmental regulatory power and the purchase of new public lands. Moreover, the federal conservation workforce is not as big as it used to be.

Even in its somewhat diminished state, however, the federal government remains the largest single employer in the world of environmental careers. Over 200,000 people worked in full-time, permanent positions for federal

Table 1. Federal Government Employment at Environmental and Conservation Agencies (2003) (All Employment/Full-Time, Permanent Employment).

Agency	September 1997	March 2003	Change
Army Corps	36,955/33,497	34,654/26,549	−6%/−21%
National Parks	21,333/14,843	21,528/11,647	None/−21%
EPA	18,393/16,592	18,652/16,475	+1%/None
Energy	17,075/16,542	15,782/15,194	−8%/−8%
NRCS	13,247/11,081	12,835/11,427	−3%/+3%
NOAA	12,095/11,491	12,396/11,716	+2%/+2%
BLM	10,611/8,722	11,161/9,773	+5%/+12%
USGS	10,519/8,695	9,954/8,113	−5%/−7%
Fish and Wildlife	7,964/6,622	9,671/8,277	+21%/+25%
Forest Service	40,760/28,778	35,002/30,007	−14%/+4%
TOTAL	188,952/156,863	181,635/149,178	−4%/−4.5%

environmental and conservation agencies in 2004, most of them at the small collection of well-known agencies detailed in Table 1.

The sheer size and authority of the government, coupled with the huge amounts of money it spends on state and local government, nonprofit groups, and private contractors, guarantee a prominent role for Washington on every sustainability issue. It's the U.S. government, after all, that owns and runs the national forests, wildlife refuges, parks, and other protected areas. Congress ultimately controls the direction of national policy on air, water, climate, toxics, agriculture, and energy issues. Federal priorities in environmental science dominate research funding for scholars and graduate students. And, in a global world, it's important to note that only Washington can commit the nation to critical international environmental and trade agreements.

Finally, it makes sense to examine the federal government's environmental workforce at length because much of the nation's public and private eco-employment has grown up in reaction or as a complement to the federal role in environmental protection and natural resource management. Understanding the federal workforce is a window into environmental career opportunities everywhere.

- Environmental Protection Agency

Through rule making and regulatory enforcement, the Environmental Protection Agency (EPA) assures that business, industry, and government agencies comply with the National Environmental Policy Act, Clean Air Act, Clean Water Act, Safe Drinking Water Act, Resource Conservation and Recovery Act (for hazardous waste management), Toxic Substances Control Act, and Pollution Prevention Act, among others. In this role, the agency often sets the standards that become our national environmental goals. Since much of this authority for on-the-ground pollution control work has been delegated to the fifty states, the EPA's role is often one of management and oversight.

In addition to responsibility for enforcing major laws, the EPA is also charged with running the "Superfund" program to clean up the nation's most toxic sites, and it has major responsibilities for environmental education, environmental justice, and brownfields redevelopment. Finally, the EPA conducts and supports scientific research.

Although the agency can be thought of as a "national cop on the environmental beat," it has shown a strong interest in additional methods for improving environmental quality. A 2004 EPA report entitled *Innovating for Better Environmental Results* reviews a long list of initiatives involving experimental technologies (such as nanotechnology), integrated environmental management systems, "sector strategies" that target the unique environmental concerns of specific industries, voluntary programs, and market-based incentives for pollution prevention. The search for innovative and less costly methods that produce improved results will certainly continue under both Republican and Democratic administrations.

Environmental scientists, engineers, and lawyers make up the vital core of the EPA workforce, for obvious reasons, but that has been changing. Information technology specialists are in very high demand now, as are economists, businesspeople, financial analysts, and talented managers for complex public-private partnerships.

- National Park Service

One of the nation's best-loved agencies, the National Park Service operates hundreds of national parks, from well-known treasures like Yosemite, Grand Canyon, and Yellowstone, to tiny properties that attract only a handful of visitors. Because many parks are simultaneously nature reserves, sites for scientific research, and outdoor recreation areas for millions of people, pressure is always on park professionals to balance public access with the need to protect the natural bounty that inspired us to protect these areas in the first place. Maintaining that balance has been made even more difficult in recent years by the need to deal with a serious backlog of needed improvements that will require millions of additional dollars beyond current budgets.

The core of the National Park Service workforce is its cadre of park rangers, a growing number of whom are unfortunately required to handle law enforcement issues along with environmental education and conservation science. Although the familiar park rangers are perennially overburdened and underpaid, the service still finds plenty of applicants for its jobs, and the appeal of a career in the National Parks remains strong.

- ## U.S. Fish and Wildlife Service

The U.S. Fish and Wildlife Service (USFWS) maintains a national network of 542 wildlife refuges and 3,000 small waterfowl nesting areas totaling ninety-six million acres—the result of a huge national investment in habitat protection. The USFWS is also in charge of a fisheries system that contains seventy hatcheries and a variety of laboratories. It carries the primary responsibility for enforcing the Endangered Species Act—one of the nation's most important, and most controversial, ecological laws. This role makes the agency a frequent target of lawsuits from environmental activists, state and local government, business interests, and property rights advocates.

The USFWS is a big employer of fish and wildlife biologists and technicians, both directly and through grants and contracts with government and academic partners. Over the last decade, the service has become a strong voice for ecosystems management approaches to wildlife protection and an important supporter of the emerging field of conservation biology. In fact, many offices of the USFWS have made a conscious shift towards employing more scientists with the management-focused education advocated by the Society for Conservation Biology, at the expense of more traditionally trained biological specialists.

- ## U.S. Geological Survey

The U.S. Geological Survey (USGS) is a nonregulatory, independent science and information agency with nearly 400 offices across every state, and also in some foreign countries as well. Perhaps best known as "the nation's mapmakers" and for its expertise on mineral resources and natural hazard risks like earth-quakes and floods, the USGS also provides data, analysis, and original research on water resources and biological diversity.

The USGS houses and updates a network of remarkably complex Geographic Information System (GIS) databases that are supported by satellite imagery, remote sensing, and people on the ground. Policy makers and scientists depend heavily on USGS for unbiased data to guide environmental decisions.

Not surprisingly, the agency's workforce is overwhelmingly technical and scientific, with earth scientists of various stripes dominating the payroll. A look at the agency's structure shows departments for mapping, water, and geological and biological resources, with the water unit employing the greatest number of professionals. Strong GIS skills are an absolute necessity for new recruits to be competitive.

- Bureau of Land Management

Often invisible to people in the eastern half of the country, the Bureau of Land Management (BLM) is huge, well known, and sometimes bitterly attacked in the western states. The bureau manages 262 million acres—nearly one-eighth of the land in the United States—and 300 million additional acres of "subsurface mineral resources" as well.

Perhaps more than any public land agency, the BLM struggles to find a balance among ecological protection, outdoor recreation, economic uses (grazing, forestry, mining, and drilling), and preservation of archaeological and historical sites. Overcoming a reputation as the bureau of "logging and mining," BLM leadership has worked hard over the last fifteen years to establish ecological protection as a central priority. Current agency issues include management of off-road vehicle use, dealing with nonnative plant and animal species, responding to a rancorous debate about possible expansion of mining and drilling exploration, and handling land use conflicts in areas where BLM land abuts sprawling western cities and towns and/or Native American communities.

Although "range managers" are the historical professional core of the BLM, hiring over the last decade shows an increase in cultural resource specialists, biologists, ecologists, water resource managers, land use planners, and outdoor recreation professionals.

- U.S. Department of Agriculture Forest Service

The U.S. Department of Agriculture's Forest Service employs 30,000 people—more than any other environmental agency. The Forest Service is also hiring more new permanent and temporary workers than anyone else. These people are needed to manage the National Forest System, an extensive network of 600 ranger districts in 155 national forests and 20 grasslands covering 191 million acres—an area the size of Texas. The agency is also involved in assisting private and urban forests, and even has a unit for international forestry.

Although logging and forest sales are still critically important, they have declined dramatically as a priority for the public forests, while ecological and watershed protection and outdoor recreation have increased. In fact, some Forest Service leaders have suggested that the agency's future may look more like the National Park Service or the U.S. Fish and Wildlife Service, both in mission and staffing.

Managing forest fires is a central concern for the Forest Service, and policy debates are fierce about how best to deal with the issue. A series of disastrous fire seasons have put the issue front and center, and fire-related hiring (including postfire restoration) is quite high.

Just as the NPS has its park rangers, the USFWS its biologists, and the BLM its range managers, the dominant job titles at the Forest Service have traditionally been those of forester and forest technician. Hiring in these professions remains strong, although the profession of forestry is more ecological than in the past and the daily activities of foresters are somewhat different than they were a generation ago. The service is also seeking ecologists, hydrologists, planners, outdoor recreation specialists, and information technology professionals.

- Natural Resources Conservation Service

Also part of the Department of Agriculture, the Natural Resources Conservation Service (NRCS) evolved from the former Soil Conservation Service and inherited its central mission—the protection and restoration of agricultural lands, watersheds, and communities. Although all twenty-first-century conservation agencies are dedicated to achieving results through partnerships, the NRCS has the longest history of meaningful involvement with its central stakeholder groups—farmers and rural communities. Because NRCS is charged with conservation of resources that are not owned by the federal government, it has developed deep connections with local people through hundreds of county offices that offer technical assistance and shared science.

In keeping with its mission, the NRCS employs a large number of people with training in agriculturally related fields, such as soil science, agronomy, agricultural economics, and watershed management. Education in these fields has evolved to include increased attention to managerial and communication issues, in response to employment signals from agencies like the NRCS.

- ## U.S. Department of Energy

The public face of the U.S. Department of Energy's (DOE) "environmental" responsibility has been focused on the past. The agency is spending tens of billions of dollars to clean up sites contaminated by over forty years of atomic weapons production and has been relentlessly seeking a place to dispose of radioactive waste from nuclear power plants. Although this is work that must be conducted (with a mission that provides lucrative contracts for many environmental firms), forward-thinking environmental professionals will probably be more interested in the DOE's policy, scientific, and technical work related to better energy use, alternate energy sources, and reduction of greenhouse gas emissions.

The department's network of twenty-four national laboratories and technology centers, for example, houses world-class facilities where over 30,000 engineers and scientists perform leading-edge research. The Energy Efficiency and Renewable Energy division—though smaller than it should be—is a fertile place for the study and promotion of hydrogen, fuel cell, solar, wind, and biomass energy sources. And even those who are concerned about fossil fuel dependence will find major scientific, engineering, and social scientific research opportunities designed to make these energy choices less environmentally damaging.

Even more than most federal agencies, the DOE pursues its mission through private contractors and formal relationships with university researchers. Therefore, many of the agency's direct employees are contract managers and administrative personnel.

- ## U.S. Army Corps of Engineers

Many people are surprised to find the U.S. Army Corps of Engineers (USACE) listed as an environmental agency, given its lead role in the past: damming, dredging, channeling, and filling of so many American rivers and wetlands.

But the USACE has serious pro-environment responsibilities, and a scientific, engineering, and management staff to match. The Everglades restoration project, for example, is largely led by USACE managers, as are innovative projects to remove dams and allow rivers to run free again. And the USACE is responsible for managing the nation's "404" permitting program for the protection and restoration of wetlands.

It's no coincidence that the agency is called a corps of "engineers." Environmental, civil, geological, and hydrological engineers dominate the agency, and engineering projects remain "job one." Still, the USACE requires a significant number of people with ecological (especially wetlands) backgrounds, as well as attorneys, project managers, designers, planners, public financing administrators, and professionals with strong negotiation capabilities.

- ## National Oceanic and Atmospheric Administration

The National Oceanic and Atmospheric Administration (NOAA) is better known by its component agencies such as the National Weather Service, National Ocean Service, and National Marine Fisheries Service. As issues such as global climate change, coastal zone management, fisheries depletion, ocean pollution, and coral reef destruction have risen closer to the top of the environmental priority list, NOAA has become something of a rising star and an employer of choice for many creative environmental professionals.

Emerging from an independent scientific focus, much like the USGS, NOAA is incorporating aspects of the EPA's regulatory culture and accepting resource management responsibilities that draw on the experiences of the land management agencies we've discussed. The combination of a marine focus, strong science, regulatory enforcement, and resource management makes NOAA a unique place to work. In addition, the agency's charge to protect the oceans and atmosphere requires it, by necessity, to take a global view. People with international perspectives and experience are welcome and needed, and this will only increase.

In keeping with its name, and with the relatively large size of the Weather Service compared to other NOAA divisions, the agency hires a large number of meteorologists and atmospheric scientists. Marine scientists, fisheries biologists, oceanographers, coastal zone managers, and GIS specialists are also well represented at the agency.

There are certainly other agencies with major environmental programs and responsibilities, the most intriguing of which may be the Department of Defense (DoD). The DoD is spending billions to clean up contaminated military bases and restore them to vibrant health. Although Congress's willingness to fund this responsibility varies from year to year, the problem is not going away, and many communities are depending upon creatively redesigned military bases for economic development now that the bases are gone. Perhaps even more interesting than remediation and restoration of closed bases, however, is the challenge of creating sustainable communities at existing ones. Military bases are small cities, and they include parks, forests, grasslands, lakes and streams, transportation networks, power plants, housing, water and sewage systems, solid waste management strategies, and so forth. More than one observer has suggested that these federally owned and operated communities could become "eco-villages," sending a strong and tangible message to the civilian world about what is possible when visionary leadership and plenty of money are combined.

In the Department of the Interior, the Bureau of Reclamation and the Bureau of Indian Affairs have compelling work opportunities, especially for those interested in tackling problems related to water quantity, quality, and distribution among competing interests in the American West. The Department of Transportation generates environmental work related to highway construction and maintenance, public transit, and even bicycle and pedestrian activity. Finally, the Department of Justice, the Department of Housing and Urban Development, the National Science Foundation, and the nation's public health agencies all deserve attention.

A new source of federal environmental employment is emerging from the war on terrorism and the creation of the Department of Homeland Security (DHS). The DHS has identified numerous terrorist targets that could generate substantial environmental impacts, including food and water supplies, power plants, the chemical industry, power stations, liquid natural gas storage facilities, and hazardous or radioactive waste sites. Professional expertise is needed to control vulnerability and to ensure quick, effective emergency response if the worst should happen. Fortunately, environmental professionals have proven experience in both prevention and response.

People who make it their business to track federal environmental jobs have been a bit confused and have had a tough job predicting trends over the last several years. Discussions of the issue have a strong "on the one hand, on the other hand" flavor. It is difficult to respond to straightforward questions about the growth or decline of federal eco-jobs with equally straightforward answers.

Nonetheless, examination of employment statistics at selected agencies between 1997 and 2003 leads to several useful conclusions. Overall, permanent employment at the ten key eco-agencies declined by 4 percent over the six-year period. The Fish and Wildlife Service (with an increase of 25 percent) and the Bureau of Land Management (with an increase of 12 percent) fared best, while the Army Corps of Engineers and the National Park Service suffered most (with a decrease of 21 percent). Most remaining eco-agencies had stagnant employment during the period and remain in a low-growth state today. Two important players—the Department of Transportation and the Department of Defense—certainly spend large amounts of money related to the environment, but direct eco-employment at these agencies is much harder to determine.

Gross employment figures mask two important trends for new entrants seeking careers in the federal government. The first is that while government hiring may be somewhat flat, many jobs have simply been transferred or "outsourced" to private contractors like CH2M Hill, Tetra Tech, Booz Allen Hamilton, and hundreds of others. People from these firms often work side by side with their government counterparts for months or even years so that it becomes increasingly difficult to make any meaningful differentiation on the ground. It's often the case that an agency may be spending significantly more money on environmental work, while actually employing fewer full-time government workers.

The second trend is that the federal environmental workforce will soon begin retiring in huge numbers. Over 40 percent of the total federal workforce is already over the age of fifty, and that means a whole generation will be leaving the job ranks very soon.

The retirement wave will hit environmental agencies especially hard, and some observers are calling the expected losses a crisis. A 2004 report entitled *Federal Natural Resources Agencies Confront an Aging Workforce and Challenges to Their Future Roles* (Renewable Natural Resources Foundation) reported that "over 50 percent of the Senior Executive Service members of the Department of the Interior (DOI), USDA Forest Service and Environmental Protection Agency will retire by 2007. Within the same period, DOI will lose 61 percent of its program managers, the Forest Service will lose 49 percent of its foresters, and EPA will lose 30 percent of its environmental specialists." Similar percentages of toxicologists, entomologists, soil scientists, chemists, and engineers will soon be eligible to retire across the entire spectrum of agencies.

Surging retirements mean that jobs are on the way for a new generation of federal government eco-workers. There is debate, however, about how many and how soon. That's because the wave of retirements is being met by equally strong counterpressures. For one thing, a roller coaster economy and stock market appear to be encouraging government workers who are eligible for retirement to stay in their jobs for a few more years, pushing the average age of workers even higher and reducing the number of new jobs for younger professionals. In addition, there is a continuing desire by many to shrink the size of the federal government workforce by cutting budgets, sending current federal responsibilities to state and local government, replacing permanent workers with part-time workers, and outsourcing even more responsibilities to private contractors and consultants. The result is that in a recent fiscal year, federal environmental agencies hired only 7,200 new, permanent, full-time workers to replace the 8,000 who left.

The tide may already be turning, however. An Earth Day 2004 review of job postings at the federal website www.usajobs.com found 1,600 permanent job openings at twelve environmentally related agencies, including 542 positions at the Forest Service and over 100 jobs each at the National Park Service, Natural Resources Conservation Service, U.S. Fish and Wildlife Service, and National Oceanic and Atmospheric Administration. The Environmental Protection Agency and the U.S. Geological Survey (with fifty and twenty-nine job postings, respectively) brought up the rear of the federal jobs on parade that day. Savvy job seekers visit usajobs.com regularly to check trends and track which fields are most in demand.

In addition, an ECO survey of selected hiring managers in the summer of 2004 revealed a very high level of optimism about the number of new jobs arriving soon and a strong desire to hire younger people with the twenty-first-century skills (especially in technology) that are becoming so critical today.

And they are putting their money where their mouths are. The number of environmentally related internships, fellowships, scholarships, educational positions, and seasonal jobs is increasing at many agencies. Agency recruiters provide detailed information about these offerings on their websites and through organizations like the Environmental Careers Organization, Student Conservation Association, and Oak Ridge Institute for Science and Education.

Table 2. Salary Table. 2004 General Schedule. Incorporating a 2.70 Percent General Increase. Effective January 2004. Annual Rates by Grade and Step.

Grade	Step 1	Step 2	Step 3	Step 4	Step 5
GS-1	$15,625	$16,146	$16,666	$17,183	$17,703
GS-2	17,568	17,985	18,567	19,060	19,274
GS-3	19,168	19,807	20,446	21,085	21,724
GS-4	21,518	22,235	22,952	23,669	24,386
GS-5	24,075	24,878	25,681	26,484	27,287
GS-6	26,836	27,731	28,626	29,521	30,416
GS-7	29,821	30,815	31,809	32,803	33,797
GS-8	33,026	34,127	35,228	36,329	37,430
GS-9	36,478	37,694	38,910	40,126	41,342
GS-10	40,171	41,510	42,849	44,188	45,527
GS-11	44,136	45,607	47,078	48,549	50,020
GS-12	52,899	54,662	56,425	58,188	59,951
GS-13	62,905	65,002	67,099	69,196	71,293
GS-14	74,335	76,813	79,291	81,769	84,247
GS-15	87,439	90,354	93,269	96,184	99,099

Those interested in federal government careers will find that competitive pay is part of the package. A large part of the federal government's environmental workforce earns between $35,000 and $75,000 per year (see Table 2), but exceptional benefits, opportunities for promotion, relative job security, and challenging work projects make federal work more attractive than base salaries alone would suggest. The federal government usually pays better than state or local government agencies for the same type of work and is often competitive with the private sector as well. Federal agencies pay substantially better than most nonprofit environmental groups.

Finally, no discussion of future environmental job prospects with the federal government would be honest without a reference to the massive federal deficit and the looming expenses for entitlements like Social Security and Medicare that await us in the future. Although the majority of Americans may not view them as such, nearly all environmental protection and conservation

Step 6	Step 7	Step 8	Step 9	Step 10	Within-Grade Amounts
$18,009	$18,521	$19,039	$19,060	$19,543	Varies
19,841	20,408	20,975	21,542	22,109	Varies
22,363	23,002	23,641	24,280	24,919	$639
25,103	25,820	26,537	27,254	27,971	717
28,090	28,893	29,696	30,499	31,302	803
31,311	32,206	33,101	33,996	34,891	895
34,791	35,785	36,779	37,773	38,767	994
38,531	39,632	40,733	41,834	42,935	1,101
42,558	43,774	44,990	46,206	47,422	1,216
46,866	48,205	49,544	50,883	52,222	1,339
51,491	52,962	54,433	55,904	57,375	1,471
61,714	63,477	65,240	67,003	68,766	1,763
73,390	75,487	77,584	79,681	81,778	2,097
86,725	89,203	91,681	94,159	96,637	2,478
102,014	104,929	107,844	110,759	113,674	2,915

agencies are "discretionary" items in the federal budget and are often slated for cuts or no-growth budgets when national security, debt, entitlements, and tax cuts eat away at federal funds. Environmental professionals have a responsibility to speak out in defense of the programs and agencies that protect our natural resources.

- State Government

While the federal government eco-workforce is big, at around 200,000 permanent, full-time employees, state government is much bigger. In fact, the fifty states employed over 270,000 environmental and conservation workers in 2004. When one considers that state governments have the same aging workforce as the federal government and that many within Congress would like to send even more eco-authority to the states, landing a job in state government begins to emerge as a great career choice.

25

Those with an interest in state government careers should familiarize themselves with the structure and responsibilities of state agencies. Most states have patterned their environmental structure after the federal government, creating separate agencies for:

1. Pollution control and prevention
2. Fisheries and wildlife protection
3. Agriculture and food safety
4. Parks and recreation
5. Water resources
6. Public health
7. Energy
8. Community and economic development
9. Coastal zone management (when applicable)
10. Planning
11. Emergency services

Recently, however, some states have begun to tear apart old structures that had grown up piecemeal in response to federal mandates or economic interests. Innovators are putting together more integrated institutions that are organized around concepts like ecosystems management, watershed protection, or even sustainable communities. Oregon, Maryland, California, Massachusetts, Minnesota, Wisconsin, and Washington are among the leaders in redesigning agency structures that allow professionals to treat ecological, economic, and social concerns as mutually reinforcing goals—not competing interests.

This state-level innovation can be clearly seen in the creation and support of markets for renewable energy, like solar, wind, biomass, small hydro, and geothermal. Through methods like renewable energy funds, renewable portfolio standards for utilities, tax incentives, green pricing programs, and direct renewable energy purchases, state governments are fueling job creation that shows up in the energy industry. Over half of the states are actively engaged in increasing the percentage of energy garnered from green sources.

Although state governments are essential to the nation's environmental protection strategy, they were hit hard from 2001 through 2003 by declining revenues, big deficits, and budget cuts. It's remarkable that the states still spent $15.1 billion on environmental work in 2003, down just a sliver from 2002's all-time high of $15.35 billion. This positive result came with a significant caveat, however. The percentage of state revenue that came from the federal

Table 3. State Government Expenditures by Environmental Category

State Government Budget Categories		Total Expenditures
Water	Water Resource	$2,186,940,159
	Water Quality	1,439,087,967
	Drinking Water	576,505,872
	Marine & Coastal	429,659,056
	Subtotal	4,632,193,054
Land Management	Forestry	1,597,786,882
	Land Management	1,090,775,125
	Soil Conservation	378,869,448
	Mining Reclamation	348,707,294
	Pesticides Control	225,225,047
	Geological Survey	179,089,909
	Subtotal	3,820,453,705
Fish & Wildlife		2,835,858,428
Waste Management	Hazardous Waste	1,581,965,318
	Solid Waste	939,234,944
	Nuclear Waste	49,099,909
	Subtotal	2,570,300,191
Air Quality		1,279,960,353
TOTAL		15,138,765,731

government in 2003 was 33 percent—higher than it's ever been since tracking started in 1986. However, state environmental spending made up only 1.4 percent of the average state's budget, the lowest in nearly twenty years. Fortunately, state managers report that the worst is over and that environmental agencies actually fared better than many other programs during the economic downturn.

So where are the jobs in the state government sector? An excellent way to track this employment is "follow the money." Because the structure of state government is so often similar to the federal government, a career seeker can use Table 3 (from Environmental Council of the States, or ECOS) to identify issue areas that have received more—and less—budget attention from state officials. By finding the comparable federal agency in the text above, one can identify trends, professions, and demands that are likely to be reflected at the state level as well. For example, the ECOS expenditures chart shows that state governments collectively spent $1.6 billion on forestry departments in 2003.

These departments are comparable to the USDA Forest Service, and the issues and personnel needs are largely the same. Consider cross-referencing water, air, and waste issues with the earlier discussion of the Environmental Protection Agency, land management with the BLM, marine issues with NOAA, fisheries and wildlife concerns with the U.S. Fish and Wildlife Service, and so forth.

As mentioned earlier, state government jobs often pay less—sometimes significantly less—than their federal and private-sector counterparts. For this reason, state agencies have a slightly higher turnover rate, as talented people gain experience on the state payroll on their way to a better paying job elsewhere. However, strong benefits packages and fewer layoffs should be considered on the plus side of working for state government.

The ECOS website (www.sso.org/ecos) is the perfect portal to launch your exploration of state government careers, as the Environmental Council of the States is an excellent source of information on state government responsibilities, policy trends and innovations, hiring, and current job availabilities.

• Local Government

There are more than 70,000 local governments in the United States, including nearly 20,000 municipal governments, over 16,000 towns and townships, 3,000 counties, and roughly 31,000 "special districts." The Environmental Careers Organization estimates that roughly 8 percent of all local government employees are directly involved in environmental and conservation work—over 480,000 people. This is roughly equivalent to the federal and state government environmental workforce combined. Clearly, local government is a big part of where the action is when it comes to environmental careers.

Special districts deserve a special mention, because they are often overlooked by job seekers. These employers are frequently found in areas such as recycling and solid waste management, electricity generation and distribution, water and wastewater treatment, public transportation, management of conservation and park land, regional planning, and air quality management. Because special districts involve concentrated groups of professionals focused on a single activity and because they often have their own revenue streams from service fees (instead of general taxes), these entities are often positive engines of creative change. Cities with positive sustainability records, like the

Portland, Oregon, metropolitan region, are achieving many of their transit and planning successes through special districts.

Work at the local level is "hands-on" environmental protection action. This is where recycling bins are picked up and the contents separated and sold. It's where water and wastewater are treated, land use plans are written and monitored, street trees are planted and trimmed, storm water runoff systems are maintained, community gardens are managed, and permits are awarded for new homes and businesses. Local government employees keep the buses running, install bike paths, run outdoor recreation programs, and act as first responders in the case of environmental emergencies like chemical spills. The local level is where the proverbial rubber meets the road.

An interesting characteristic of local environmental work is that many jobs do not require advanced degrees, or sometimes even a college education. Technical degrees, certifications, or perhaps just a high school background may prove sufficient—with the real learning coming in the school of practical experience. Perhaps because of this, local government workers with extensive and detailed knowledge developed through years "in the trenches" sometimes do not get the credit and respect they deserve as serious environmental professionals.

By necessity, local government environmental employment is growing more rapidly in places where the population is expanding. We find significant numbers of new jobs, for example, in Georgia, Florida, Texas, Virginia, the Rocky Mountain West, Arizona, and many parts of California. Growth is less lively in the rural Midwest and in much of the Northeast.

Another curious thing about local government jobs is that, although there are hundreds of thousands of them, new positions can sometimes be difficult to find and more difficult still to obtain. On reflection, it's not hard to see why. A given town or city may have only a small handful of environmental professionals, and good new jobs get snapped up fast. Local governments are often strapped for cash as well and may simply choose to add additional responsibilities to the local land use planning staff instead of hiring new people when new problems or opportunities arise. This also means that local governments are increasingly looking for candidates with higher education.

Over the last twenty-five years, local governments have formed closer ties with the environmental industry, and many jobs that were government positions in the past have migrated to the private sector. Your local solid waste crews, for example, are just as likely to get a paycheck from a firm like Waste

Management or Allied Waste as they are from the local public utility department. The same trend exists in local electric utility work, and even water utilities are creating partnerships with companies like U.S. Filter, further blurring the lines between public and private employment. Although most of these arrangements fall short of outright privatization (where firms actually own assets that were previously in public hands), those with a strong interest in local environmental work may find themselves in a business career.

Perhaps the most important of all, local leaders and professionals are a very big part of the movement for sustainable communities in the United States. It's at the local level that connections between ecological and social health are most immediately experienced and where the results of policy experiments can be quickly assessed and improved upon. As cataloged by the U.S. branch of the International Council of Local Environmental Initiatives (www.iclei.org), local governments are at the forefront of this work.

To get a more detailed understanding of the scope of local government environmental jobs, the best place to begin is literally at home. Take a walk through the government listings in your local phone book, identify agencies that seem interesting, and begin talking with key managers and professionals.

• Nonprofit Organizations

No one knows how many environmental nonprofit groups there are in the United States. Estimates range as low as 4,000 and as high as 10,000 or more. There has been a large (and welcome) growth in small, local, grassroots groups in the last ten years. Many of these groups, however, have no staff. The number of nonprofit groups of interest to career seekers drops precipitously if one eliminates all volunteer groups.

Although no comprehensive census has been done, we do know that there are:

- Over 1,400 land trusts;
- Over 2,000 water-related groups;
- 1,450 nature centers;
- Hundreds of chapters of national and regional groups, such as Sierra Club, Audubon Society, and Izaak Walton League;
- More than 2,000 environmental justice groups;
- Thousands of small neighborhood and community groups devoted to environmental improvement;

- Hundreds of animal rescue and rehabilitation groups;
- 90 aquariums; and
- Over 1,000 student groups on college campuses.

The list can go on to include church projects, garden clubs, scout troops, rails-to-trails organizations, summer camps, museums, and more.

The structure of staffed nonprofit organizations, wherever they are, is remarkably similar. Usually, there is an executive director who manages the organization, as well as an administrative assistant. In many cases, that's the whole staff. Larger organizations will also have a fundraising and membership department, program staff for major activities, finance and accounting personnel, education and communications people, lobbyists and attorneys, a volunteer coordinator, and project coordinators for grant-funded initiatives. Core staff are supplemented by interns and volunteers.

The largest environmental nonprofit group is certainly The Nature Conservancy, which employs over 2,500 people all over the world, as well as many seasonal employees and interns. Most groups are considerably smaller.

Thirty of the better-known groups in the nonprofit world are American Farmland Trust, American Forests, American Lung Association, American Rivers, Appalachian Mountain Club, Audubon Society, Clean Water Action, Conservation Foundation, Defenders of Wildlife, Earth Island Institute, Eco Trust, Environmental Defense, Friends of the Earth, Greenpeace, Izaak Walton League, Land Trust Alliance, League of Conservation Voters, National Parks and Conservation Association, National Wildlife Federation, Natural Resources Defense Council, Public Citizen, Public Interest Research Group, Rails-to-Trails Conservancy, Resources for the Future, Sierra Club, Trust for Public Land, Wilderness Society, World Resources Institute, World Wildlife Fund, and Worldwatch Institute.

Taken together, these thirty influential groups employ perhaps 4,000 people, probably less. Environmental Defense (ED), for instance, is one of the best-known environmental groups in the world. The paid staff at ED is just over 160, and it is a big environmental group. It's not surprising that full-time staff positions at well-known environmental nonprofits are among the most competitive of all environmental career offerings.

The environmental nonprofit world is headquartered in a few major cities. Certainly, Washington, D.C., is the geographic center of nonprofit environmentalism. Many of the groups above have a heavy concentration of staff people in the nation's capital, as well as its Maryland and Virginia suburbs. New York

City, Boston, San Francisco, Chicago, Denver, and Seattle are also popular centers for large nonprofits.

Nonprofit work in recent years has had a few consistent trends. First, management standards have increased. Boards are demanding that directors and managers have (or get) strong management and leadership skills. Passion and commitment are not enough. Second, revenue-generating ability is crucial. A lot of people can come up with good ideas, but fewer can make them pay through fee-for-service programs, membership, sales of supporting goods and materials, and fundraising. Those who can are in demand. Third, nonprofits are learning to work together. Funders (and the public) are asking that nonprofit managers learn to form effective alliances and collaborations. Fourth, nonprofit environmental groups are learning to diversify, creating an environmental movement that appeals to all Americans, regardless of class, race, and ethnicity. Finally, environmentalism is returning to the grassroots with the understanding that a concerned, involved, informed, and politically savvy citizenry is essential for environmental success.

- Schools

There are over 14,000 school systems in the United States, employing 3.5 million full-time employees and more than 1 million part-timers. After accounting for custodians, principals, bus drivers, administrators, secretaries and so forth, the lion's share of this dedicated workforce is, of course, teachers.

It's impossible to make an accurate count of how many teachers could be categorized as "environmental" educators, although it is clear that more and more districts are finding it difficult to afford specialists of any kind. Studies have shown that most designated environmental education is given to science teachers, but years of anecdotal information make clear that the "environmental" educators in most schools are simply those who choose to bring environmental learning into the curriculum.

Employment trends for teachers in many states are quite encouraging. The number of teachers needed is certainly growing, and those with expertise in science and math are particularly needed.

Private Sector

Quick! Name the top five environmental services companies in the nation. How about waste management firms? Global solar and wind energy leaders? Which grocery chain sells more organic food than anyone else?

If you drew a blank on these questions, you're in good company. While most people easily recognize the EPA and Sierra Club, both the leaders and the laggards of the environmental business world are largely invisible. Even many environmental professionals are unaware of just how much work is done by the private sector—and how many jobs are created as a result.

Business employers can be found in three broad categories:

- Traditional "environmental industry" companies
- Environment, health, and safety departments inside regulated corporations
- Sustainable businesses

The Environmental Industry

The easiest way to describe the "environmental industry" is through its major component parts. Water and wastewater treatment is by far the largest, followed by solid waste management, air quality, site remediation, and hazardous waste. Although no firm statistics exist, the most conservative industry estimate puts the total at 400,000 people—including a large number of environmental engineers and technicians—on 2003 revenues of more than $220 billion in the United States alone.

Let's start with water. The dream of many in the water business is a movement for virtual privatization of municipal water systems across North America, especially those in major metropolitan areas. Any widespread change from public to private ownership of water utilities would result in one of the largest industries in the world, with a revenue stream in the hundreds of billions of dollars. High-profile controversies in Atlanta and New Orleans, as well as an extremely skeptical public, make this scenario extremely unlikely. Still, there is plenty of work in the private water business involving a wide spectrum of different public-private partnership models.

The $92 billion water industry in the United States draws 35 percent of its revenues from water utilities, 32 percent from wastewater treatment works, and a significant portion from selling instruments, lab services, equipment, chemicals, consulting, and "contract operations." French conglomerate Veolia, owners of the biggest U.S. firm, U.S. Filter, is a primary industry leader. This list also includes United Water, Thames Water (a British giant), and a handful of other large entities, as well as dozens of smaller players.

A key concern (and hope) for the water industry is what's known as "the funding gap." *Environmental Business Journal* reports that studies from the EPA, the American Water Works Association, and the American Metropolitan Sewerage Association "estimate that water/wastewater systems will require anywhere from $250 billion to $1 trillion in new investment above existing levels in the next 20 years." Clearly, this national need could be a big growth opportunity for the industry.

Unfortunately, state governments are in no financial position to deal with the need, and the federal government seems less than interested in providing serious leadership. Beyond the simple need to repair pipes and plants, there are other drivers creating water-related business opportunities. Federal "Total Maximum Daily Load" (TMDL) requirements and "Phase II" storm water management rules both require local and state officials to meet standards that add to the nation's already strong water laws. Finally, the international market for water and wastewater improvements is exploding.

The result of all of this action is a water industry that is large and fairly healthy—employing over 200,000 people—but that continually feels perched on the edge of something that could be much, much bigger, eventually on a par with the energy, food, and transportation industries as an engine of economic growth and development.

As might be expected, the water industry employs large numbers of civil engineers, an army of water and wastewater technicians, and substantial numbers of hydrologists, surveyors, planners, financing specialists, project managers, and community involvement professionals.

Although smaller than the water/wastewater business, solid waste management is a huge enterprise, generating a conservatively estimated $65 billion a year. According to *Waste Age* magazine, the top 100 firms in the field alone employed over 154,000 people in 2003—many of them at small companies with workforces of around 100 to 150 employees. The sector is overwhelmingly dominated, however, by a short list of leaders that includes Waste Management, Allied Waste Industries, Onyx North America, Republic Services, PSC, Covanta Energy, Safety-Kleen, Waste Connections, Casella Waste Systems, and Stericycle. Don't blink, though—the leader board seems to change regularly in this field.

One trend that the solid waste field shares with the water business is an ongoing process of mergers, acquisitions, bankruptcies, and breakups. Although the overall direction seems to be moving toward greater consolidation, with fewer and fewer giants controlling more and more market share, it's probably more accurate to say that the entire environmental industry is searching for the right balance between size and flexibility to deliver environmental goods and services.

There was a time when it appeared that the environmental business would be about boundless growth, not planned consolidation. The industry came into existence at a historical moment when government regulations, spending, and industry compliance needs were growing and changing at a rapid pace. Hundreds of billions of dollars were spent on new cleanup activities, some of them site-related, but most requiring expensive technical fixes and ongoing maintenance requirements. Although government priorities shifted from general pollution control (1970s), to hazardous waste remediation (1980s), to air pollution issues (early 1990s), to water concerns (late 1990s), it appeared that there would always be something new. However, those days seem long gone, and new drivers requiring government or industry expenditures have been in short supply.

Numbers tell the story. In the 1970s, growth was 9 to 11 percent annually. The 1980s saw 10 percent annual returns in the first half of the decade, rising to 12 percent and peaking at 15 percent in 1988. Throughout the 1990s, growth was flat to moderate, ranging from a low of around 2 percent to as high as 6 percent. Today, overall growth rates are quite low, and some of the industry's past revenue sources are even declining. Although increases in productivity, cost savings, and increased prices can make a difference, eventually something's got to give. In lieu of new markets, companies turn to mergers and acquisitions for growth, settle into a specialized "niche," hunker down until the next big thing, or go out of business.

This dynamic is being played out now in the environmental design, engineering, and consulting world. A 2004 report on the top 200 environmental firms in *Engineering News Record* (www.enr.com) showed that annual environmental revenues for these leaders alone was $31.4 billion—4 percent less than the 2003 total and the second straight year of decline. The national customer base for these firms was 34.5 percent private businesses, 32.7 percent federal government agencies, and 32.9 percent state and local government clients. Although many leaders were optimistic about the chances for their own companies, no one was expecting big spending increases from either public or private sources any time soon.

As it has for years, water/wastewater issues (42 percent) and hazardous/nuclear waste concerns (42 percent) dominated environmental spending for the leading environmental consulting, engineering, and design firms. The remaining 16 percent of business was distributed among environmental science (5.7 percent), environmental management (5.2 percent), and air quality (3.3 percent). On an extremely depressing note, "sustainability" didn't yet show up enough to be a separate category, and few executives at environmental service firms mentioned it as a concern on which people were willing to spend money.

Nonetheless, these firms are hiring people now, because $33 billion is still a lot of money and the work is labor-intensive. To prove the point, ECO picked four firms randomly from ENR's top 200 (Versar, Arcadis, Tetra Tech, and CDM) and found 785 active job listings online on a day in 2004 from those four companies alone—overwhelmingly for engineering, scientific, technical, project management, and information technology positions.

For the record, the 2004 list of environmental service firms was topped by U.S. Filter, Bechtel, CH2M Hill, URS, Washington Group, MWH, The Shaw Group, Parsons, Fluor Corporation, and Tetra Tech. Check out the *Engineering News Record* website for the complete annual rankings, which are used by many environmental job seekers as an online job search database.

The market areas about which environmental industry types seem least optimistic are remediation, hazardous waste, and air quality—collectively a $50 billion revenue source. It would be nice to think that these sluggish marketplaces are simply the result of a job well done—sites are cleaned up, air quality is exceptional, and hazardous waste is under control. And, in the case of hazardous waste, there is some truth to that. The effectiveness of monitoring, shipping, treating, and disposing of hazardous waste has improved dramatically, the cost has dropped, and the number of people who are needed has declined. Lacking expanded regulation, the only driver left to create business in the field is a hard-charging economy with increased production and related growth in hazardous waste. As CH2M Hill's senior vice president David Rosenblum told *Engineering News Record*, "An awful lot of traditional hazardous waste work depends on the economy. You don't see any change in the volume of work other than for economic reasons."

In the remediation world, the problems are more political. There are still many sites to be cleaned up, including many under the jurisdiction of the Department of Defense and the Department of Energy, and a large number are designated as Superfund sites. And there are some known problems that would generate big markets if regulators moved in. Our rocket fuel disposal problem, for example, could create the need for action that has been estimated as high as $40 billion. Moreover, there are as many as 400,000 "brownfields" in the nation—contaminated and abandoned properties that are too suspect to build on but not hazardous enough to meet the standards for government cleanup. What's lacking, of course, is money. Without such funds, environmental firms will look for work elsewhere, unless they can team up with real estate development companies on brownfields projects that leave behind profitable commercial and industrial properties.

Finally, the $24 billion air quality market is an odd case. The big bump the industry received from the 1990 Clean Air Act Amendments wore off years ago (with a dramatically improved air pollution control system left behind). Many observers thought that concern over global climate change would turn air quality into a hot place to be for environmental firms, whether through the Kyoto Protocol or something like it. But it hasn't happened—at least not in the United States. Nor has the government acted to take the worst power plants off-line, raise auto efficiency standards, or require more stringent control of mercury emissions. Without such changes, the air market will remain stale.

Wherever one works in environmental industry—water, solid waste, consulting, hazardous waste, remediation, or other opportunities—there is one positive factor that is almost universally true: salary levels are higher than local and state government work, and usually higher than comparable federal government positions as well. Competition is intense for talented people among the top firms, and companies are willing to pay for people who bring them business and keep customers coming back. The negative side of the pay equation lies in the ease and alacrity with which executives lay people off when business turns downward. Still, environmental industry salaries are extremely competitive and suggest that one need not take a big pay cut to do something good for the earth.

- ## Corporate Environmental, Health, and Safety (EHS)

The corporate response to the regulatory flurry of 1970 to 1990 was the creation and expansion of "EHS" departments, staffed by professionals who were well-versed in the intricacies of environmental regulations, Occupational Health and Safety Administration (OSHA) regulations, and related laws. All larger regulated companies have well-developed EHS offices that either manage environmental compliance and remediation directly or develop contractual relationships with environmental industry firms like those just discussed. The EHS staff is usually small and often made up of people with backgrounds in environmental engineering, industrial hygiene, chemistry, and environmental science, supplemented by legal assistance and clerical help. For many years, the establishment and orderly operation of an effective EHS department was the primary corporate response to environmental concerns, and the success was measured primarily by three yardsticks: secure all necessary permits, comply with the law, and keep costs as low as possible.

Brad Allenby, former environmental vice president at telecommunications giant AT&T, is one of many corporate executives who feels that the "EHS" department may have outlived its usefulness and is probably a doubtful place from which to launch an environmental career. He's only partly joking when he says that "if you care about the environment, don't work in the environmental department."

Allenby is one of the leaders in the emerging field of industrial ecology, a radically different approach that calls for serious changes in both process engineering and human systems, and only a nominal focus on complying with regulations. "In business, everyone wants to reduce overhead—expenses that don't generate any income," Allenby says. "And, most EHS departments are overhead, because they're structured around compliance with the law—a standard that's usually so low it can't create value for the company except in a hard-to-define risk management way. That kind of EHS isn't making money. It's only preventing the potential loss of money. It's hard to verify the return on investment (ROI) when a dollar is budgeted for EHS, and that makes it a hard sell for senior managers who evaluate everything else based on ROI."

Although there are still jobs for people with traditional EHS training and experience, business leadership seems to be leaning toward managers who can work with departments like product design, manufacturing, distribution, marketing, and sales to reform industrial and business practices in a way that achieves measurable business results. The challenge is to achieve those results while improving and protecting the environment, worker safety, and public health.

Allenby stresses that moving away from the EHS model also means emphasizing different skills—especially management and leadership. "As difficult as the technical stuff is," he notes, "forging agreement with other managers to spend serious money on something really new is much harder."

• Sustainable Businesses

Which businesses should be counted as "sustainable" firms? Perhaps the day will come when profit-making enterprises are required by law to submit audited financial statements against the "triple bottom line" of ecological health, economic security, and social justice, resolving the definition question for all of us. Since that day is not here yet, and since this is a book about environmental careers, perhaps we can define sustainable businesses through a narrow ecological lens as "companies whose products and services consciously aim to reduce environmental damage while meeting a human economic demand."

If the traditional "waste and water" environmental industry discussed above is characterized by slow growth from a massive base, then "green business" is all about rapid—even phenomenal—growth from a starting place that was close to zero. Environmentally focused goods and services have had literally nowhere to go but up for the last several years. Six business areas lead the "green business" world: wind energy, solar power, fuel cells, organic food, certified forest products, and ecotourism.

Each one of these industries has its own version of the same "good news/bad news" story—only the statistics and the company names change. Consider the following examples.

Information from trade associations representing wind power like American Wind Energy Association, organic food like Organic Trade Association, and ecotourism businesses such as the International Ecotourism Society all claim that these "green" products and services are "the fastest growing part" of their industries. And, they're right. What sometimes goes unsaid, however, is that even after several years of growth that consistently hits 20 percent or more per year (good news), all three of these industries continue to represent the tiniest fraction of their total industry's sales (bad news).

Organic food is a good case in point. A 2003 report from the U.S. Department of Agriculture notes that "consumer demand rose throughout the 1990s—20 percent or more annually—and pace has continued. Organic products . . . now account for approximately 1 to 2 percent of total food sales in the United States." So, while it's true (as the firm Datamonitor reports) that "the U.S. organic market is projected to reach a value of $30.7 billion by 2007" with a compound annual growth rate over 21 percent from 1997 to 2007—it will still be less than 3 percent of total U.S. food sales, at best.

A similar good news/bad news story relates to employment growth for those farmers who want to go green. The Rodale Institute hopes that the United States might have 100,000 certified organic farmers by the year 2013. That would equal 5 percent of the nation's two million farmers—a big increase for today's total of only 13,000 certified organic farmers.

Numbers like these are good enough, however, for a company like Whole Foods Markets—the nation's first certified organic supermarket chain and the largest "green" grocer in the nation. While many conventional chains are struggling, Whole Foods is expanding existing stores, opening new ones, hiring hundreds of people, and producing high-quality profits at the same time. Executives at the firm claim that demand for organics is hard to predict based on the past and that the market share of American food sales that go to organics may reach 10, 15, or even 25 percent much faster than anyone expects.

Wind and solar power advocates make similar claims for their industries. From 1999 to 2003, the global wind power industry has grown (measured by increased "installed capacity online") at an average pace of more than 26 percent—with the leading markets being Germany, the United States, Spain, India, and Austria, and strong growth in the United Kingdom and Japan. At the end of that growth, wind power still only accounts for 0.05 percent of world electricity supply.

Solar photovoltaics tell a similar tale, with even more spectacular growth, but an even smaller percentage at the end of the story. World solar photovoltaic market installations soared to 574 megawatts in 2003, a 34 percent increase over 2002. Germany alone grew by 76 percent, and Japan, Germany, and the United States together accounted for 75 percent of the world market. World solar cell production rose 40 percent in 2003—a huge one-year increase for any industry. In 2004, however, the total contribution of photovoltaics to electricity supply was less than 0.5 percent.

Unlike the wind business, which is focused exclusively on one product (turbines), the solar energy industry includes a wide variety of specialty applications beyond photovoltaics including solar hot water heating systems, transpired solar collectors that preheat ventilation air, passive solar designs, and "daylighting" systems that reduce the need for electric power sources. Tracking sales and employment growth in these niche areas is more difficult than in the more commodity-driven area of photovoltaic sales.

If the solar and wind industries are different, however, they have one very important trait in common—domination by a small group of multinational corporations, most of them from outside of the United States. The wind business is led by Vestas Wind Systems, GE Wind Energy, Enercon, and Gamesa. Only one is in the United States. The overwhelming leaders in the photovoltaic cell manufacturing business are Sharp Electronics and Kyocera, both Japanese companies. Other important names include Shell Solar, GE Energy, BP Solar, Mitsubishi, Sanyo, Siemens, Isofoton, Photowatt, and RWE Solar.

It's clear that we're not kicking the fossil fuel habit anytime soon. But, people are buying green energy and the growth is intriguing. As one small example among many, five U.S. utilities sold more $120 million of renewable kilowatt hours from green pricing programs in 2003, led by Austin Energy ($289 million), Portland, Oregon, General Electric ($189 million), and followed by Sacramento Municipal Utility District, PacifiCorp, and Xcel Energy.

The holy grail of green energy, of course, is the development of the so-called hydrogen economy, with its promise of boundless energy and no polluting emissions to foul the air or contribute to global warming. No mining, no drilling—an environmentalist's dream.

If we are ever to move beyond the age of oil, coal, natural gas, and nuclear power, harnessing hydrogen will likely be the answer. Getting there will not be easy, due to the fact that fossil fuels themselves—especially natural gas—are needed to produce hydrogen in a usable form that can then power cars or generate electricity. The expense is high in the auto example—more than ten times for a hydrogen fuel cell car as for an internal combustion engine.

Building or retrofitting an energy infrastructure to complement a shift to hydrogen will cost a lot of cash as well. According to Associated Press reporter Mark Johnson, "General Motors Company has estimated that it would cost $11.7 billion to build 6,500 hydrogen fuel stations in 100 metropolitan areas throughout the United States, and 5,200 more on national highways."

Clearly, there are hundreds of billions of dollars to be made for those who solve the hydrogen problem, and investment money from government and industry is pouring into research and development efforts, as well as actual hydrogen power projects. The Associated Press reported in April 2004 that the United States is spending $200 million this year on hydrogen research, compared with $260 million in Japan. Canada will spend $400 million, and the European Union will invest more than $2.5 billion over the next several years. These amounts are in addition to President George W. Bush's much-touted pledge to spend $1.7 billion on hydrogen research and development over the next five years—$350 million of which was put into the scientific engineering pipeline in 2004. The world's automakers and energy companies all have large-scale hydrogen programs as well.

A key player in the development of a hydrogen economy is the fuel cell industry. Fuel cells are battery-like devices that combine hydrogen and oxygen to produce electricity without combustion. Although most talk about fuel cells is centered on future hopes, the present isn't bad either. The Electric Power Research Institute has estimated that U.S. fuel cell sales in 2004 will reach nearly 300 megawatts—compared to far less than 100 megawatts in 2003—and estimates for 2005 run over 600 megawatts. Fuel cells are widely used in a variety of energy applications that don't require a grid connection.

Nearly everyone predicts a fuel cell business explosion that will dwarf today's numbers. And that explosion is likely to take place outside of the United States first, according to David Jollie, editor of *Fuel Cell Today*, based in London. "Anywhere there is not an established grid structure or infrastructure there will be more opportunity to get in there (with fuel cells)," Jollie told the Associated Press. "They'll look to leap frog the old centralized distribution structure." Countries like China and India—with over two billion people, but without large-scale electricity distribution networks—have high hopes for hydrogen. (Coal and natural gas, however, will almost certainly be used in producing it.) Japan, with its high energy demands and oil prices, will likely join them.

What does all this mean for renewable energy job seekers? First, there is a huge demand for scientists and engineers with specialized expertise in wind, solar, hydrogen, and fuel cell basics and technology. Talented people with focused training will find no shortage of job opportunities in both basic research and applied development. Second, the industry requires an infusion of business talent—managers, sales and marketing people, financial analysts and accountants, contract lawyers, tax experts, and related professionals. Finally, installations on the ground will mean jobs for technicians and maintenance people who can keep things up and running.

The numbers are very large. An April 2004 report from the University of California's Renewable and Appropriate Energy Laboratory (http://socrates.berkeley.edu/~rael/papers.html) found that renewable energy (including biomass) created jobs at a rate that was many times higher than fossil fuels, when manufacturing, installation, operations, and maintenance were taken into account. PricewaterhouseCoopers has predicted that by 2013, the North American fuel cell industry will provide jobs for 108,000 people in the manufacturing sector alone.

There is more to tell, of course. The green business niche is growing quickly in fields like architecture, landscaping, construction, forest products, energy conservation, investment banking, consumer products, and tourism. Ultimately, our goal must be to blur the line between "environmental" work and the rest of the nation's workforce.

Building Your
Environmental Career

Understanding global and national issues is critically important for all environmental professionals. To build and maintain a successful career, however, requires much more. Mastering the seven basic steps below will put a person ahead of the crowd in today's ultra-competitive environmental marketplace.

Seven Steps to a Great Environmental Career

1. Know yourself
2. Get focused
3. Know what's going on in the world around you
4. Start growing your career network
5. Get the targeted skills and experience you need
6. Master the job search basics: resumes, cover letters, research, and interviews
7. Be a great performer

Step One: Know Yourself

Many professionals spend over half of their waking hours at work. Think about that for a moment. With so much riding on it, it's essential that your job be a good fit.

Fortunately, there are many wonderful resources available to environmental job seekers in the form of career advisors and coaches, books, websites, and classes. One tool that many people have found useful is the Myers-Briggs Type Inventory (MBTI). Based on an extensive questionnaire and professional follow-up, the MBTI provides valuable clues to help job seekers select work that maximizes their unique personality. Most college career offices offer the Myers-Briggs analysis.

Whether you use Myers-Briggs, some other tool, or simply quiet reflection, self-knowledge is critical for career advancement.

Step Two: Get Focused

Respected career coach Barbara Sher tapped into a rich vein of anxiety when she titled a book, *I Could Do Anything . . . If I Only Knew What It Was*. Narrowing your job search focus is perhaps the most difficult task on the way to landing a great environmental job. Without a clear sense of where you want to end up, it's almost impossible to create a plan of action to get there.

Getting focused does not mean putting yourself in a job straightjacket that will eliminate all other options forever. Nor does it mean nailing down every detail. Although some people can confidently say "I want to be an Oceanographer II at the Coastal Services Center of the National Oceanic and Atmospheric Administration in Charleston, South Carolina, for the next 3.5 years," most job seekers are much more general in their job hopes.

In today's interdisciplinary environmental world, one powerful tool for narrowing your career focus is to ask, "What do I want to create in the world?" Directing attention to the results you want to be a part of is useful for a number of reasons. Most importantly, this focus taps into a powerful trend in environmental work away from process and toward dramatic results.

Defining your job focus around achieving positive, dramatic, and measurable changes in the world has other advantages. It demonstrates to others that you are not looking for "a job," but rather for an effective platform from which you can make a difference. It opens you to an interdisciplinary network of professionals who are committed to creating the same change that you are.

Developing career direction based on the change you want to create is also perfect for those who have a wide array of different interests and therefore have a difficult time making a decision that seems to narrow future options. Most environmental issues are linked to one another. A person who focuses on air quality will soon be addressing transportation, energy production, and more. In a very real sense, it doesn't matter where you start. It only matters that you get started.

Step Three: Know What's Going On in the World Around You

Without self-knowledge and focus, it is impossible to be proactive, increasing the possibility that a job seeker will simply respond to whatever comes along. Unfortunately, as everyone eventually learns, the employers of the world do not select people for jobs primarily on the basis of the applicant's hopes and dreams. Employers have agendas of their own, and the assets of each job seeker are viewed through the prism of those needs.

To be a master of the job search process, you'll need knowledge of the trends and pressures that are guiding job selection decisions. Generally speaking, this information falls into four major categories:

1. *Economic and societal "mega trends."* The globalization of the world's economy is one example of a mega trend.
2. *Issue-specific trends.* People with an interest in global climate change, for instance, need information about leading-edge policy, action, funding, and research.
3. *Sector-specific trends.* Sectors are broad employment categories, such as state governments generally or the land trust community.
4. *Employer-specific trends.* Detailed information about organizations of interest to you.

Staying on top of this information might seem daunting—even impossible. If you have defined a career focus around making a particular change in the world, however, the task becomes much easier.

Suppose, for example, your current passion is water rights issues in the American West. To keep a finger on the pulse of all four types of trends, you need only tap into a well-organized career network (see step four) for information on crucial journals to read, websites to visit, organizations to track, conferences to attend, and innovative leaders to seek out.

Step Four: Grow and Maintain Your Career Network

Most jobs are never advertised, and many of those that are advertised are eventually filled by people who are already in the employer's network. With this in mind, career advisors universally agree that all serious job seekers need a vibrant network of professional colleagues, friends, neighbors, relatives, and acquaintances to be successful. Years of surveys and studies support this commonsense observation.

Even so, "networking" has something of a bad name. The word has become associated with soulless opportunism and as verification of the cynical belief that career advancement is based on "who you know, not what you know." Perhaps there are times when networking lived up to that nasty reputation. At its best, however, networking is a positive process through which people who are creatively engaged in making a difference support others who share their passion.

Whether a job seeker views networking as a necessary but unpleasant evil or as an exciting opportunity to create lasting friendships with fascinating people, it's critically important for all job seekers to become skilled networkers. Effective career networking includes:

- *Reciprocity.* You must give more than you take.
- *Referrals are essential.* If you identify a person that you would like to meet, either for direct career reasons such as job leads or simply to share ideas, the most effective course of action is to seek a referral from someone respected by that person.
- *Quality, not quantity.* The strength of a person's network is measured by whether it has the right people in it for his or her needs. It's not just a big Rolodex.
- *Get out there.* Attend meetings. Send notes to speakers and writers whose work you admired. Pick up the phone and introduce yourself.
- *Little things mean a lot.* A well-timed "thank you" note, for instance, is helpful.
- *Stay in touch.* Most of us are better at starting a relationship than maintaining it, but maintaining relationships is one of the most important aspects of effective networking.

Step Five: Get the Skills and Experience That You Need

Every job announcement has a list of required qualifications. Although sometimes these qualifications are incomplete, misleading, or inaccurate, the focus on needed skills and experience is there for an obvious reason. Employers are looking for talented people with proven ability to get the job done.

No amount of "job search technique" can substitute for simply having the right mix of talents that employers seek. Employers seem to agree with a workforce assessment study prepared for the EPA in 1999. That study suggested a balance of ten essential abilities.

Top Ten Skills for Twenty-First-Century Environmental Professionals

1. Communication ability (speaking, writing, visual, listening)
2. Collaboration ability
3. Creativity and innovation
4. Broad environmental science understanding
5. Analytical and critical thinking/problem-solving ability
6. A positive attitude/willingness to work hard
7. Information technology skills, including geographic information systems
8. Leadership ability
9. Occupation-specific skills
10. "Customer" orientation (focused on the needs of stakeholders)

Step Six: Master the Job Search Basics

There are dozens of great books and websites about resume and cover letter writing, interviewing techniques, locating job listings, using the Internet, and so forth. Professional advisors at college career centers are also extremely competent at teaching these skills. Take advantage of these dedicated professionals!

Step Seven: Be a Great Performer, Leave a Legacy

With strong self-awareness, results-oriented focus, knowledge of trends, a growing career network, a balanced set of skills, and mastery of killer job search tactics, it won't be difficult to land a job, even in a difficult economy. Once you do, always remember: with each passing year, career advancement is mostly about stellar performance.

Conversations with the Experts

WORKING THE ISSUES

1. AGRICULTURE AND FOOD SECURITY

Dr. Fred Kirschenmann.
Photo used courtesy
Successful Farming magazine.

A CONVERSATION WITH
Fred Kirschenmann

Having spent his life working in the field, literally, as well as on the marketing and sales side of agricultural production, Fred Kirschenmann is keenly aware that modern industrial agricultural practices jeopardize long-term agricultural stability.

He's hopeful that redesigning of our agricultural system can make us more aware of the "most basic details of our own food production" and also help counter a production system driven mostly by economic forces. "We need to think about where we want to go in the future with agriculture," Kirschenmann says. "The principles of sustainability must become actualized in ways that will work for farms and farm families."

Kirschenmann is director of the Leopold Center for Sustainable Agriculture at Iowa State University. He previously managed the 3,500-acre certified organic Kirschenmann Family Farms in North Dakota and was president of Farm Verified Organic, a private organic certification agency. Kirschenmann obtained a Ph.D. in philosophy from the University of Chicago and is the author of numerous articles and book chapters dealing with ethics and agriculture.

What Is the Issue?

As we enter the twenty-first century, modern agriculture stands at a cross-roads. We've had unprecedented success in increasing the yields of some commodity crops and animal species, largely due to the introduction of hybrid crops and modern technology during the "Green Revolution" in the 1960s. But agriculture is now directly affected by some extraordinary new challenges. These include fossil fuel depletion, environmental degradation, climate change, biodiversity loss, population growth, persistent poverty, and an unprecedented explosion of infectious diseases. Some believe these challenges can be met by developing a new generation of technologies. Others believe that only a new paradigm for food, fiber, and energy production can surmount the challenges ahead.

• Environmental Degradation

Masae Shiyomi and Hiroshi Koizumi argue that two emerging challenges associated with modern agriculture—the depletion of fossil fuels and environmental degradation—will force agriculture to make some dramatic changes in the near future. Modern industrial agriculture is enormously dependent on fossil fuel. We need it to run farm equipment and produce fertilizers and pesticides. But we are currently pumping out of the second half of the oil barrel, making this resource increasingly costly.

At the same time, degradation associated with agriculture is becoming increasingly intolerable, and it is unlikely we can continue ignoring the costs that agricultural wastes impose on others and on the environment. There are now at least fifty hypoxic zones on the planet, which are areas where excess algae growth has robbed coastal waters of enough oxygen to support most marine animals. All of these areas are associated with nutrient pollution from industrial agriculture regions.

At the same time, environmental change is posing new challenges to agriculture. For example, climate change impacts on agriculture are becoming apparent. A recent Iowa State University study projects that the Upper Mississippi River Basin is likely to see precipitation increases of 21 percent by 2050, causing surface runoff increases of up to 51 percent. Such runoff would dramatically increase soil erosion and nutrient pollution.

According to Lester R. Brown, writing in *Eco-Economy: Building an Economy for the Earth*, some 36 percent of the world's cropland is already losing topsoil at rates that are undermining productivity. In short, modern agriculture is possible because of abundant, cheap natural resources, like fossil fuels and virgin soils that are now in a state of depletion, and because of natural sinks that absorb wastes, although these sinks are now saturated. Neither can be relied upon to subsidize production systems indefinitely.

Additionally, an unprecedented explosion of infectious diseases, mostly caused by ecological impact, presents further unique challenges because of concentrated animal agriculture systems. According to writer Mark Walters, writing in *Orion*, "Crammed into factory farms, pigs and poultry live in the equivalent of a Petri dish," which is an ideal environment for disease.

- ## Biodiversity Loss and Genetic Uniformity

In the interest of maximizing efficiency, virtually every modern agricultural production system has become specialized to produce just one crop. These specialized systems, called monocultures, significantly decrease biodiversity. By some estimates, traditional agriculture once used 80,000 plant varieties for food production. Today, 80 to 90 percent of the world's calories are produced from just ten or twenty crops. Animal diversity has been similarly compromised. Increasingly narrow production specifications are to blame and have seriously compromised the gene pool. Ecologist Norman Myers even suggests that wheat could become an endangered species "because of a protracted breeding trend toward genetic uniformity" that has caused wheat to lose "the great bulk of its populations." Consequently, wild strains and genetic diversity are almost nonexistent.

Unfortunately, U.S. political efforts have not encouraged the maintenance of broad-based genetic seed stocks. The recently debated energy bill, for example, contained provisions supporting greater production of ethanol, a product based on corn, a staple of monoculture farming. This is just one example of government policy that is completely at odds with sound ecological thinking.

- ## Pest Management

Increasingly, agriculturalists have recognized that managing farms in an industrial manner produces numerous problems. For example, Joe Lewis, pest management specialist with the U.S. Department of Agriculture (USDA), points out that applying pesticides to solve pest problems is not sustainable. This "solution" ignores basic ecological principles and dynamics that cause pest emergence and also leaves systems ripe for resurgence. Since all organisms develop defenses against threats to their existence, we then apply even more lethal pesticides to surviving pests, often resulting in harmful consequences to other organisms. All too often, the solution becomes part of the problem.

At the same time, applying fertilizers to maintain crop yields also masks poor soil management, causing erosion and compromising soil quality. Inevitably, we must then increase fertilizer use as counteraction. Furthermore, concentrating livestock in one place produces excess manure that can't be economically transported to fields where manure would be a welcome source of fertility. Applying inappropriately high rates in one location and then purchasing synthetic fertilizers in another leads to increased nutrient runoff, which contributes to impaired water systems.

• Economic Costs

The farm sector of agriculture is in crisis even without counting the costs of environmental damage, which are either subsidized by taxpayers or simply never accounted for at all. In fact, if it weren't for huge government subsidies, most farmers couldn't afford to pay their bills. Ever since the 1980s government subsidies have comprised a significant portion of net farm income. In 1993, for example, government subsidies accounted for 143 percent of Iowa farmers' net farm income.

These economic situations have multiple causes, but chief among them is the cost of input-dependent industrial agriculture, market structures that leave farmers without power to negotiate cost-of-production prices, and public policies that favor agribusiness corporations—all issues that significantly impact small farmers. Furthermore, these situations may be among the principal reasons that few farmers have used direct marketing strategies or have transitioned to producing higher-value products such as salad greens or herbs or high-value meat products such as grass-fed organic instead of continuing to produce undifferentiated commodities such as feed corn.

• Bioterrorism

These problems are compounded by potential threats of bioterrorism to food security. According to David Orr, a society fed by a few "mega farms" is much more vulnerable to disruption than a society with many smaller and widely dispersed farms. Farms that rely on long-distance transport must guard supplies carefully, but often the military's capability to do so is ecologically costly, contributing to more vulnerability. As Orr writes, "In short, no society that relies on distant food, energy, and materials sources or heroic feats of technology can be secured indefinitely." His observations coincide with Yale sociologist Charles Perrow's views. Perrow points out that in large, complex, and tightly coupled systems, accidents that generally occur in any system will always become catastrophes because of vulnerability. This is not the case in smaller, dispersed systems.

- Persistent Poverty and Food Insecurity

Now more than ever, the global community is recognizing and demanding that food is a basic human right. Worldwide, we are faced with continued, often dramatic, population growth, mostly in regions beset with poverty. The United Nations estimates that the world's population will reach 9.3 billion by the year 2050. Most of the additional 3.2 billion people will live in poor rural areas in the developing world. World poverty rates are increasing at 100 million people per decade. Nearly half of the planet's 6 billion people now live on less than two dollars a day.

Hunger and famine are directly associated with population growth and persistent poverty, although they are caused by many more, enormously complex factors. For example, Barry Berak points out that Africa's paralyzing debt, sorry infrastructure, depleted soil, meager exports, bad government, and ethnic neglect all contribute to these problems. Simply inventing costly new technologies to increase productivity in the developed world, where overproduction already forces farmers to accept profits well below the cost of production, will not solve this problem. More creative agricultural solutions are needed in order to feed the increasing world population and equitably distribute the world food supply.

How Are Environmental Professionals Approaching the Issue?

A recent National Academy of Sciences report, *Frontiers in Agricultural Research: Food, Health, Environment and Communities*, recommended that the USDA shift the emphasis of its $2 billion annual research budget from increasing food and fiber production to promoting environmentally sound farming alternatives, higher quality of life in rural communities, better diet and health, increased food safety, and a softer impact of globalization on U.S. farming. This report also suggested that scientists see a need to shift to more ecologically based agricultural approaches. But currently few scientists are trained to address food and farming issues from this perspective, and few farmers have practical knowledge in applying this approach. Several initiatives might help us make progress toward a new revolution (see Box 1-1).

• Developing Ecologically Based Agriculture

In discussions about how we should approach agriculture's future, agriculturalists Shiyomi and Koizumi ask, "Is it possible to replace current technologies based on fossil fuel energy with proper interactions operating between crops, livestock, and other organisms to enhance agricultural production?" If the answer is yes, then we could redesign farm systems using a systems-based approach that would create a diverse habitat with many species that complement each other, instead of attempting to increase productivity by introducing technologies that force a single species to yield more.

Several ecologists have begun to show an interest in this type of production agriculture. University of Minnesota ecology professor David Tilman, for example, suggests we recapture and build upon the ecological principles that guided agricultural production prior to the "Green Revolution," a term coined by agronomist Norman Borlaug in the 1960s. While Borlaug bred high-yielding strains of dwarf wheat and rice to be grown in intense monocultures, Tilman suggests that we use newly accumulated information in ecology and evolutionary biology to redesign future systems. In addition to knowledge of ecological

BOX 1-1

STRATEGIES TOWARD A GREEN REVOLUTION IN AGRICULTURE

- Developing Ecologically Based Agriculture
- Returning Wastes to Crop Production
- Exploring New Ways to Produce Crops
- Finding Economic Solutions
- Developing New Policies

ORGANIC FARMER

AT A GLANCE

Employment:

Over 17,000 certified organic farmers

Demand:

Rising rapidly

Breakdown:

Private sector, 100 percent

Trends:

• Organic production is growing more rapidly in traditional commodity crops than in local produce, creating a need for farmers to raise and sell products both in local markets and to organic food manufacturers.

• Passage of U.S. Department of Agriculture guidelines for organic product certification has encouraged more traditional growers to enter this field.

Salary:

Although potentially lucrative, average incomes are often low. Organic farmers earn as little as $15,000, and most usually average $30,000 annually. Some organic farmers' earnings can exceed $100,000 depending on size of operations and specialities.

JOB DESCRIPTION

Daily work in organic farming varies widely depending on the type, scale, and diversity of products. Organic dairy farms or livestock operations, for example, require different skills than small vegetable outfits, which also differ from skills needed at orchards, large commodity farms, or operations producing niche products like peppermint for tea. "There's no such thing as a typical day," says Nicole Vitello, owner of Manic Organics, a family farm in Portsmouth, Rhode Island. "Small organic farmers do a little bit of everything, including seeding, hoeing, weeding, running irrigation, harvesting, and going to market." Business management is also critical for small farms. Vitello believes that savvy marketing, cost control, and financial record-keeping are almost as essential as work in the field.

The U.S. Department of Agriculture's strict organic certification guidelines require farmers to rigorously verify the results of natural solutions. "Farmers must provide thorough descriptions of crop and livestock production practices and keep records tracking products from field to point-of-sale," says Donald Burgett of the Organic Farming Research Foundation. "This ensures organic fields are actually free of pesticides, nonorganic fertilizers, and other contaminants." Like all organic farmers, in order to meet such certification standards, Vitello searches for natural solutions when dealing with diseases or pests that infest her crops, rather than applying pesticides.

Beyond production, organic farmers spend time learning about methods to increase crop yields and tap new markets. As an owner-operator, Vitello reports that turning profits often means long hours for herself and her family, instead of hiring paid help. "It's a challenge," she says, "but also an opportunity to learn what Mother Nature is all about."

GETTING STARTED

• Organic farmers must understand agriculture, economics, and farm machinery. Farm work experience is necessary.

• Many farms offer internships introducing people to agricultural techniques and farming through the Collaborative Regional Alliance for Farmer Training (www.csalearningcenter.org/craft).

• Grower associations exist in most states and metro regions. With notice, most farmers are glad to share wisdom about their particular niche.

• Knowledge of the USDA National Organic Program is essential. Visit www.ams. usda.gov/nop/indexIE.htm.

ADVICE FROM THE PROS

Klaas and Mary Howell Martens switched their 1,300-acre farm in upstate New York to organic because they could not make sufficient profits the traditional way. Today, they grow bulk products like corn, soybeans, cabbage, dried peas, and barley, which they sell to large organic food manufacturers. "The overhead for organics is huge and will sink those who don't know what they're doing and who don't start out slowly and carefully," Klaas Martens says. "I encourage people to work for a period of time in different markets to gain knowledge. That experience will reduce mistakes, which is critical because when you make farming mistakes, there's no one to cover you. When you're ready, find a specific niche and stick with it."

processes, this "greener" revolution will combine and incorporate feedbacks, disease dynamics, soil processes, and microbial ecology, ultimately using "the principles of ecology, epidemiology, evolution, microbiology, and soil science" all together.

To prove the benefits of this approach, Tilman conducted a ten-year research project comparing the productivity of corn grown according to the best industrial methods with corn grown according to alternative ecologically based practices. While the ten-year average maize yields differed by less than 1 percent and profitability was almost equivalent, the ecologically based system demonstrated many other positive advantages, such as significant increases in soil organic matter, stable nitrogen content, and 60 percent less nitrate leaching into groundwater as compared with the conventional system. (See Tilman, "The Greening of the Green Revolution," for more details.)

- Returning Wastes to Crop Production

Clearly, ecologically based agriculture has its roots in agricultural practices that are guided by the example of nature. In his seminal work, *The Ecology of Commerce*, Paul Hawken suggests that all human economies be redesigned in accordance with principles "guided by the example of nature." These principles include the recycling of all waste, switching to renewable energy, and restoring everything we use.

When waste from any activity doesn't return as food for another part of the system, that waste eventually exceeds the planet's capacity for absorption. This waste-equals-food principle is known in agriculture as the "Rule of Return." Returning wastes to soil creates humus, which produces healthy crops, whose properly composted remains improve the soil content. In integrated crop/livestock systems, for example, "waste" from crop systems (such as straw, weed seeds, broken and shrunken kernels that aren't suitable for human consumption) can be fed to livestock, while that livestock's waste (manure) can be spread on fields to restore the humus content of the soil. A truly integrated system might first extract energy from agricultural waste by feeding it into a digester and then return the refuse from the digester to the farm fields to restore soil fertility.

Overall, a new day may be dawning in both agriculture and ecology. Productive farming may be increasingly dependent on insights from ecology and evolutionary biology. Instead of being entirely dependent on manufactured

ENVIRONMENTAL CERTIFICATION SPECIALIST

AT A GLANCE

Employment:

Approximately 200 people worked for five companies in 2003.

Demand:

Increasing

Breakdown:

Private sector, 100 percent

Trends:

• While no studies document growth within this industry yet, it is expected that consumer demand for eco-certified products will increase demand for certification professionals.

• Although many corporate professionals transfer into mid-level positions, there are many entry-level positions available at this time.

Salary:

Starting salaries begin at $30,000, and median earnings are equal to approximately $55,000. Some can earn as much as $80,000 or more.

JOB DESCRIPTION

As environmental concerns grow, industries and manufacturers are making a concerted effort to develop more sustainable practices and products. To curb "green washing," independent certification companies are emerging to ensure that companies adhere to environmentally sound principles. Wendy Hall, a Certification Administrator for SmartWood, says she left the high-end corporate world to pursue a more meaningful career. "After 20 years working in corporate business, I strongly felt a need to feel good about what I was doing," she says. Like Hall, environmental certification specialists are those professionals working behind the scenes to establish corporate accountability.

Environmental certification specialists take clients, such as organic farms, timber companies, and manufacturers, through certification processes. This includes inspecting the origin of goods to be processed, records, equipment, and production and processing systems. If a company meets specific industry standards, they obtain certification. For instance, the fishing industry follows guidelines set by the Marine Stewardship Council, while manufacturers follow ISO 14000, a set of standards dictating the production of "environmentally preferable products."

If a specialist notes environmental problems during the evaluation process, the company must work with an outside consultant fix these problems within a deadline. Environmental certification specialists only work with clients during the certification process rather than on improvements. Once companies correct any problems, environmental certification specialists then reassess operations or products and often grant certification.

Although much of a specialist's time is spent working with new clients, they also address problems that already-certified companies experience. Environmental certification specialists draw from a large knowledge base and are often experts within their area of specialization. For example, forestry certification specialists have an extensive background in forestry sciences. In addition, they are well versed in sustainability standards within the industry they work for and have a strong knowledge of markets and business.

GETTING STARTED

• Most specialists have at least a master's degree in their area of expertise, but it is also important to incorporate training in environmental auditing into your studies.

• A background in marketing and business is desirable.

• Check out groups like Smart Wood (www.smartwood.org) and Green Seal (www.greenseal.org) to see what the certification process looks like and what kind of products and industry are getting certified.

ADVICE FROM THE PROS

Richard Z. Donovan, chief of forestry at the Rainforest Alliance, says that the environmental certification field provides a built-in market process that can "foster significant impact on industry." With this in mind, aspiring professionals should focus their education in forestry, agriculture, or another environmental field. This way, they may focus their efforts on improving a company's business practices in those specific industries. "Work for a business trying to make money in these fields so you can learn how it operates," he adds.

inputs, future farmers may increasingly depend on nature's wisdom. Effective conservation may increasingly depend on managing farms as habitats instead of factories. The continuing economic and environmental pitfalls of industrial agriculture have left small farmers with few options—they can get bigger, get out, or change the way they have been farming to work in harmony with natural systems and processes.

• Exploring New Ways to Produce Crops

The shift in production agriculture suggests enormous challenges and opportunities for agriculturalists, ecologists, and evolutionary biologists interested in researching and exploring ecological agriculture approaches. These professionals should work together to create a framework that can identify and implement a research agenda. Such collaboration could also identify training and resource needs for future professionals working in the sustainable agriculture field.

One thing we can do now to start implementing this agenda is to identify the world's successful ecologically based farms and use them as learning laboratories to provide both farmers and researchers with extraordinary, experience-based information about managing ecologically based systems. Takao Furuno, a farmer in southern Japan, has developed a highly innovative duck, rice, fish, and fruit farm by integrating various species into a highly productive, synergistic system. Joel Salatin in Virginia has developed a similar multispecies system suitable to his region that is equally productive.

Indeed, many conservationists, ecologists, and environmentalists have shown intense interest in this new direction in agriculture. In *The Forgotten Pollinators* (Island Press, 1996), Gary Nabhan and Steve Buchman have demonstrated habitat development's essential role in production agriculture, especially in cultivating pollinators. Dan Imhoff, with the Wild Farm Alliance, has published *Farming with the Wild* (Sierra Club Books, 2003), which describes in detail thirty-six farms nationwide where wildlife habitat management is an important ingredient of farming. As a biologist at the University of Northern Iowa and associate director of the Land Stewardship Project in Minnesota, Laura and Dana Jackson have similarly featured farms where the hard line between tame and wild has been erased to the benefit of both farming and conservation. They've also outlined ecological principles that guide farmers managing these farms in their recent book, *The Farm as Natural Habitat: Reconnecting Food Systems and Ecosystems* (Island Press, 2002).

Farmers and researchers have only begun exploring nature's rich resources for a farming and habitat future that could accomplish several goals:

1. Stem environmental damage and restore its productive capacity
2. Restore and enhance genetic and biological diversity
3. Manage pests appropriately rather than attempting to eradicate them
4. Recognize the role of local and regional production in enhancing food security

Recently, for example, a group of British scientists including Thorunn Helgason, Tim Daniell, Rebecca Husband, Alastair Fitter, and Peter Young discovered that photosynthate from healthy forest trees growing in full sunlight can travel via fungi that grow on tree roots to reach weaker, shaded trees to bolster their impoverished nutrition. Such ecological processes, they argue, are "vital components of community diversity." Their work validates the need for more research about ecological processes and nutrient management systems that ecologically based agriculture might imitate or incorporate. Only time and directed research for such approaches will determine if they can make agriculture more productive and food systems more efficient and secure.

• Finding Economic Solutions

The day of shifting environmental costs of production onto others is rapidly ending, and many business enterprises are now discovering that ventures can be designed to be both profitable and ecologically restorative. For this reason, ecologists and economists should explore ways to work together more effectively, especially with respect to agriculture.

New production and marketing models are emerging that are less costly to farmers and the environment, and they enable farmers to retain more of the value of their production. Farmers' markets, Community Supported Agriculture (CSA), and organic farming have each provided markets for a small number of producers. Numerous additional efforts are under way, even among some major food companies, to connect farmers who practice good stewardship with food customers who prefer to buy from them. However, a more fundamental policy and market restructuring will be required to address the grave ecological and economic problems facing U.S. agriculture.

Career Spotlight

Stonyfield Farm
Londonderry, New Hampshire

Sustainable agriculture is not a utopian idea for some far distant future. Firms and farmers are making it happen right now and none better than a yogurt company with 215 employees nestled in the small town of Londonderry, New Hampshire. The firm is Stonyfield Farm, and it has quietly become the nation's third-largest yogurt maker, with $150 million in annual sales from organic yogurt, frozen yogurt, ice cream, and cultured soy products. Stonyfield has proven so successful in reaching the mainstream market that the world's number-one yogurt maker (French conglomerate Group Danone) has purchased a 75 percent stake in the company.

Stonyfield Farm was founded in 1983 as a project at a family-run organic farming school. The co-owners had seven cows. In that first year, the company pushed to produce 150 cases of yogurt a year for the New England marketplace. Today, the firm churns out 350,000 cases for yogurt lovers in all fifty states—all made from natural and organic ingredients and without bovine growth hormone (rBGH).

In keeping with its sustainability mission, Stonyfield donates 10 percent of profits to education and support practices that protect the natural world. It also alerts consumers about environmental issues by placing educational information on the lids of its yogurt containers, reduces plastic consumption by using lighter-weight cups than other manufacturers, and replaces plastic lids with foil seals. After the sale, the company collects its yogurt cups and recycles them into products like toothbrushes.

Stonyfield's success is pleasing, but not surprising to Gary Hirshberg, the company's co-founder and CEO. "Consumers want delicious food made from natural ingredients by people who care about the community and the environment—and that's what we offer," he explains simply.

One of the first graduates of Hampshire College in Amherst, Massachusetts, Hirshberg worked as an activist and educator before shifting to business. He's done everything in his twenty years at Stonyfield, including milking cows, delivering the yogurt, selling to stores, and raising capital for the business. And although he doesn't mind retelling the story of the company's funky past, he's much more excited about Stonyfield's current people and future prospects.

The husband and wife team of Lisa and Erik Drake are representative of Stonyfield employees—Lisa as one of two company natural resources project managers and Erik as the Stonyfield Farm product manager. Both earned

B.S. and M.S. degrees in civil and environmental engineering, logged career stints with environmental consulting firms, and came to Stonyfield Farm because it embodied their social and environmental concerns.

Lisa Drake's job at Stonyfield Farm is to help the company further improve its already significant environmental track record. Toward that end, she carefully tracks and reports on the company's energy, water, and other natural resource uses; seeks ways to reduce greenhouse gas emissions; improves recycling efforts to reduce solid waste; and guarantees that wastewater emissions exceed government standards. "Fulfilling the mission is absolutely a team effort," she says. "In addition to managing natural resources directly, our department plays an important role in internal education, working to keep our principles on the forefront of people's minds."

As product manager for Stonyfield Farm, Erik Drake's responsibilities include developing, managing, and implementing marketing strategies and tactics for several of the company's newest product lines, including yogurt smoothies, cultured soy, and dessert products. Achieving success for these offerings in a fiercely competitive marketplace calls on Eric's skills in product strategy, developing new products, packaging, consumer promotions, and market research. To hone those skills, Eric uses knowledge gained from his environmental background, a University of Michigan M.B.A., and two years at the Johnson & Johnson Consumer Products Company.

As Stonyfield Farm moves further into its third decade, there is one part of its past that its CEO is working hard to regain. Once an all-organic company, Stonyfield began labeling itself "natural and organic" in the mid-1990s, when cost and supply issues made it difficult to ensure 100 percent organic products in a rapidly growing product line. Today, 15 percent of Stonyfield products (all in the nonfat line) have some nonorganic content, but Stonyfield Farm intends to be all-organic again by 2005—and reaching a bigger share of the nation's yogurt eaters than ever.

- Developing New Policies

Ecologists and public **policy analysts** should also explore more effective ways of working together to create sustainable agriculture systems. Public policies often are posed as a choice between protecting the environment and stimulating the economy. However, we are discovering that many win-win policies are possible. For example, the Conservation Security Program (CSP), incorporated into the 2002 Farm Bill, was crafted with strong cooperation between farm groups and conservation advocates. The CSP provides incentives for farmers to implement ecologically restorative practices on their farms and improve economic performance, in the process providing numerous public goods for current and future generations. Similar collaboration between ecologists, policy makers, and farmers could develop additional public policies with similar outcomes. Paying subsidies to farmers who preserve or restore water and soil health could replace paying subsidies to farmers to overproduce commodities they can't sell.

Career Advice from the Expert

Once scientists and farmers begin applying insights from ecology and evolutionary biology to agriculture, they will recognize that specialization is not the answer. Consolidating production and processing has created multiple problems for agriculture, and a systems approach is needed to solve those problems. Students interested in this approach should craft educational study that is holistic, broad based, and teamwork oriented.

Many colleges and universities have programs in environmental sciences or ecological studies that provide a sound beginning. They acknowledge agriculture's need for humanistic grounding by offering courses or seminars in bioethics. Combining academic background with internships or other experience will give students the best practical experience for work in the field.

For example, undergraduate students at Iowa State University (ISU), where I work, can participate in an internship program called "Life in Iowa" that offers academic credit for working with rural agricultural enterprises. ISU's curriculum for its sustainable agriculture graduate program offers a multidisciplinary approach that taps into the knowledge base from every facet of agriculture. It also offers students from other colleges, like ISU's College of Business, a minor to increase their marketability. Other universities are making similar multidisciplinary programs available to students.

Students interested in academia should realize that the future market for scholars in this field will change radically. There has been a major decline in publicly supported agricultural research. This will continue, along with an increase in special funding for various projects. Accordingly, future academicians must have grant-writing skills, which will also be invaluable for anyone wanting to work with nongovernmental organizations.

Finally, prospective farmers should be flexible in what they produce and how. They should think beyond production to marketing, as they will have to deal with people more often to understand market specifications and customer needs. By taking a more active role, these prospective professionals will realize more profits.

RESOURCES

American Farmland Trust, www.farmland.org

Center for Rural Affairs, www.cfra.org

Ecological Farming Association, www.eco-farm.org

The Food Project, www.thefoodproject.org

International Food Policy Research Institute, www.ifpri.org

The Land Institute, www.landinstitute.org

Leopold Center for Sustainable Agriculture, www.ag.iastate.edu/centers/leopold

Organic Trade Association, www.ota.com

Stonyfield Farms, www.stoneyfieldfarms.com

U.S. Department of Agriculture, www.usda.org

United Nations Food and Agriculture Organization, www.fao.org

2. AIR QUALITY

Tom Cackette in the California Air Resources Haagen'smit vehicle laboratory in El Monte, California. Source: California Air Resources Board.

A CONVERSATION WITH
Tom Cackette

As the chief deputy executive officer of the California Air Resources Board (ARB), Tom Cackette is serious about eliminating air pollution. "California leads the nation by adopting air emissions standards that are more stringent than federal regulations," he says. For this reason, the entire nation looks to the exemplary work that Cackette helps direct.

Prior to working at ARB, Cackette gained firsthand knowledge of living in the smoggiest city in the United States. He worked for the private firm Rocketdyne in Los Angeles, where he was involved in rocket engine production and test and flight performance analysis. He also worked at the EPA Motor Vehicle Emission Laboratory in various technical, management, and policy positions.

Cackette holds an M.S. degree in engineering from California State University. He speaks frequently on air quality issues and has published papers for the Society of Automotive Engineers and the Air and Waste Management Association.

What Is the Issue?

Air pollution results from the direct emission of certain chemical substances, in the form of gases or very small particles. Among gases, we can look at nitric oxides (NO_x) as an example of air pollutants that should cause concern. Nitric oxides are emitted when fuel is burned at high temperatures in the air. A problem by itself, NO_x can also react with gasoline vapors in the atmosphere to form ozone, commonly called "summertime smog." It's no secret that California has a problem with smog, and summertime ozone levels in some areas here can be twice as high as those considered safe.

At cooler times of the year, gaseous NO_x can be a different kind of problem, as it transforms into fine particles in the atmosphere, such as nitrate. Nitrates and other fine particles such as diesel soot are toxic and cause health problems and premature death. Unfortunately, in many urban areas of California, particulate matter levels are sometimes three times as high as judged safe by science and law.

Whether measured as gases, particles, or chemical combinations of both, air pollution is one of our biggest environmental problems. Most urban areas in the nation suffer from unhealthy air on one or more days per year, and many on tens or even hundreds of days annually. In California, for example, 90 percent of the population is exposed to unhealthy air at some time each year—and this is after decades of hard work and billions of public and private dollars spent.

Until recently, most air pollution concerns were relatively local or, at most, regional. We are now becoming increasingly concerned about emissions of greenhouse gasses and their relationship to global climate change, which appears to be significantly harming our welfare and the planet's ecology.

• Environmental and Economic Degradation

Air pollutants most obviously alter general visibility, inhibiting our ability to see beautiful vistas because they are blocked by gray, hazy skies. Although this may seem like a minor impact when compared to human health concerns resulting in hospital stays and premature death, it has a powerful effect on people, the environment, and the economy. The cost of the corrosive effects of air pollution can add millions of dollars to maintenance costs for buildings and other structures. Because people do not want to live in high-smog areas, the haze of pollution affects housing prices, business locations, and settlement patterns.

Crop yields are also affected by air pollution, with significant impacts. The San Joaquin Valley of California is one of the largest agricultural production areas in our country, producing 80 to 90 percent of some crops that benefit the entire nation. Research shows that summertime smog exposure directly reduces agricultural yields in this area, resulting in the loss of crops worth millions of dollars every year.

- Public Health Impacts

Poor air quality is not just a nuisance or a haze that stops people from seeing the mountains. In fact, air pollution is a serious health concern. Consider asthma. About 10 percent of young people have asthma, and air pollution is proven to trigger asthma attacks. In places with extremely poor air quality like parts of California, the number of asthma attacks connected to pollution exceeds 300,000 per year.

Air pollution has also been causally related to triggering or worsening a variety of other diseases, producing thousands of additional days of hospitalization that bring misery for the ill and the loss of thousands of work days, significantly impacting our economy. And it gets worse. Ozone can cause lung disease. Toxic air pollutants can cause cancer and birth defects. Finally, in California alone, approximately 6,500 people each year actually die prematurely as a result of air pollution, making it one of the leading killers in the state.

- Environmental Justice Concerns

While basic air pollution is spread throughout urban areas and is specific to locale, low-income communities, communities of color, and high-density communities tend to be located closer to polluting industries than middle-income communities. Therefore, they can have greater exposure to toxic air pollution. These communities also witness greater truck traffic due to industry proximity, creating even more exposure to air pollution such as diesel particulates. According to the South Coast Air Quality Measurement District, the air pollution control agency for the four counties in southern California, diesel soot or particulate accounts for approximately 70 percent of the exposure to toxic air pollutants in the state's urban areas. This pollution can cause higher concentration and risks of cancer, such that 500 in one million people are at risk in such urban areas, while in most other areas of the country, the risk is less than 10 people exposed per million.

• Population Growth, Land Use, and Transportation

Many states all over the United States are growing, especially in suburban areas. Increased populations have greater impacts on the environment, whether people are driving more vehicles, using more consumer products, or buying products whose manufacturing, production, and distribution emits pollutants. These problems affect all communities with poor air quality. Although each activity has different impacts on air quality and we are reducing emissions through targeted programs in specific industries, all reductions are partially offset by continued growth. For instance, both the number of vehicles and their use has grown due to population growth and sprawling community layout. These increases often result in continually increasing vehicle miles traveled. Affordable housing tends to be located on the fringes of urban areas, while business activities are centrally located, meaning that those who can only afford certain housing prices are again disproportionately affected.

• Emissions Sources

Emissions come from multiple sources that have controls that are not uniform, creating a problem when it comes to their regulation. The California Air Resources Board often separates emissions sources into stationary, area-wide, and mobile categories. In the state, 60 percent of emissions that cause summertime smog come from mobile sources, including cars, trucks, and construction and farming equipment. Although emissions amounts vary according to different pollutants, half of hydrocarbon emissions come from vehicle use, and about 80 percent of nitrogen oxide emissions come from cars and vehicles with heavy diesel engines. This makes vehicles the primary sources of emissions on the West Coast. Here, we have already implemented stringent regulations on cars that effectively reduce emissions. Diesel-powered vehicles, however, continue to give off significant amounts of smog-forming emissions and direct soot emissions.

Besides vehicles, the rest of the emissions contributing to poor air quality comes from stationary sources, such as consumer activities that may involve the use of solvents, paints, and personal care products, which currently contribute to about 15 percent of the pollution problem. Other sources include combustion sources like oil refineries and power plants. Although California's combustion sources have been heavily controlled through stringent emissions controls, this is not the case nationwide. On the East Coast, for example, industries burn coal and oil, producing fine particles like sulfate. These industries have fewer regulations, making them a large emissions source in that region, which certainly contrasts with the more efficient energy production facilities in California.

How Are Environmental Professionals Approaching the Issue?

- Regulating Emissions

Enacting and implementing air quality laws are important strategies in our current regulatory society. Performance standards and government regulation structurally control most activities that produce air pollution in the United States. The goal of these standards is to improve overall air quality by achieving levels of air pollution emissions that are not harmful to the public.

BOX 2-1

STRATEGIES TO IMPROVE AIR QUALITY

- Regulating Emissions
- Spurring Technological Innovation
- Generating Incentives
- Qualifying the Benefits

The Clean Air Act is one of the most successful laws ever passed by Congress and is an example of successful regulation. This law has institutionalized the idea that pollution is harmful. It has also strengthened, rather than disturbed, our economy. As a result, almost all urban air pollution has decreased significantly despite population growth, proving that achieving clean air targets is possible even though we are not yet completely pollution free.

Almost all sources of emissions in the United States have some kind of emission control requirement today, making products much cleaner now than they were before regulations were imposed. Regulations for coal-based power plants are one example of successful emissions controls. Nationwide, we are trying to reduce sulfur oxide, nitrogen oxide, and particulate matter from these sources. Strong regulations in some areas of the country have forced many power plants to use exhaust after-treatment devices to reduce smog-forming emissions. California is leading by example in this arena, as all of its plants are powered by natural gas. Particulate emissions are minimal, and the plants also use advanced controls to reduce NO_x emissions.

Regulations for oil refineries, which typically leaked emissions when petroleum was processed, refined, and transported, are another example of successful emissions controls. These regulations require refineries to have floating fuel seals on storage tanks that prevent evaporation, as well as vapor recovery systems at both ends. Other improved controls include inspection and maintenance programs to help keep refinery flanges and fittings leak free. Environmental scientists and chemical, mechanical, and **environmental engineers**, often working at air pollution control agencies, determine the feasibility of different technologies to reduce emissions, like flanges and fittings in this case. Overall, regulatory action for this specific technology has resulted in 90 percent reductions (or more), although not all processes in refineries are controlled to this extent. To ensure that these pollution regulations are effective, **environmental protection technicians** monitor and analyze emissions, while regulators enforce the controls.

• Spurring Technological Innovation

Technology itself is the leading factor in reducing emissions so that they meet health-based standards. Although there may be specific chemical concentrations to which people can be exposed without harm—like certain levels of

ENVIRONMENTAL PROTECTION TECHNICIAN

AT A GLANCE

Employment:

30,000

Demand:

Good

Breakdown:

Public sector, 60 percent

Private sector, 40 percent

Trends:

• Increases in the number of Superfund and brownfields sites will create more jobs for environmental protection technicians.

• Retirements in the next few years will open up more entry-level positions.

Salary:

Starting salaries for environmental protection technicians begin at $25,000, with median annual earnings of $36,000. Most earn between $30,000 and $45,000 with top salaries equal to $55,500.

JOB DESCRIPTION

Environmental protection technicians spend most of their time testing air, water, and soil samples to ensure compliance with local, state, and federal environmental protection laws. These professionals are also routinely called on to assist in the cleanup of contaminated land like brownfields and Superfund sites. In this aspect, they work with engineers and environmental regulators to determine the extent of damage and which procedures will work best to clean up the site.

Sarah Bates, a biologist with the U.S. Army Corps of Engineers, specializes in cleaning up military bases with shooting ranges that often contain high levels of lead contamination from ammunition left behind from firing practice. "The first thing you want to do is look around at a site and notice any obvious areas of contamination," she says. "After that, you could research and see if the site has a history of contamination. Then, you go into the sampling work." In addition, technicians must know proper cleanup procedures and environmental laws, such as the Resource Conservation and Recovery Act and the Clean Water and Clean Air Acts, to ensure their cleanup activities comply with these standards.

In the field, environmental protection technicians must be comfortable using field equipment, test kits, oxygen meters, and other devices to collect samples and determine contamination levels on a site. "Because it usually involves much soil and groundwater testing, this work often involves working outdoors," Bates adds. After collecting samples, technicians often work in laboratories, using instruments to monitor experiments, calculate and record results, and develop conclusions. They must keep detailed logs to keep track of their work activities.

At times, these technicians contract with private businesses and government agencies to manage the release of hazardous materials and conduct regular facility inspections. Likewise, private businesses also hire environmental protection technicians to manage these activities.

GETTING STARTED

• An associate's degree from a community or technical college is required for employment.
• Look for schools that offer certification in environmental management or remediation; wastewater treatment; and hazardous, toxic, and radiation waste safety training.
• Learn public speaking and technical writing, as most technicians must report their findings through oral or written reports.
• Gain strong computer skills, as most technicians work with computer-interfaced equipment on a daily basis to perform sampling work.

ADVICE FROM THE PROS

Nick Nowicki, owner of Nowicki & Associates in Washington, performs asbestos and groundwater cleanup for governmental agencies, schools, and businesses in the state. "One thing's for sure, you can't be afraid to get dirty in this field," he says. "You should also expect to experience all kinds of working and weather conditions."

ozone, for example—many toxic materials must have emissions levels set to near zero to protect public health. In areas with severe air pollution problems, turning as many pollution sources into "zero emissions" sources as possible will be the only way to ultimately achieve clean air. We need sophisticated chemical and mechanical engineers to develop proper technology to achieve these goals.

One example of technological innovation targeting zero emissions is California's low emissions vehicle (LEV) program, the ultimate goal of which is to introduce fuel cell vehicles. En route to creating and implementing such technology, the automobile industry is currently producing gasoline-fueled vehicles whose emissions are close to zero. Where a muscle car of the 1960s put out nearly two tons of smog-forming emissions during its fifteen-year lifetime, today, California's "clean cars" (which are also available in New York, Massachusetts, Vermont, and Maine) emit less than ten pounds of emissions during their lifetime.

Examples of super-ultra-clean cars include the 2004 models of the Honda Accord, Ford Focus, Toyota Camry, several Volvo models, and one BMW model. These cars are the best performing and least polluting conventional cars ever produced. Their pollution levels are reduced by 99 percent compared to cars without emission controls, and they are powerful and reliable. Going even further, hybrid electric vehicles such as the Toyota Prius and Honda Civic and Insight have even lower carbon emissions and get better mileage.

The next step in requiring low emissions vehicles is to put similar regulations in place for diesel trucks, SUVs, and other heavy equipment. Right now, mechanical and chemical engineers are working to retrofit existing diesel engines with devices that can collect and burn off diesel particulate or soot. Within this decade, both new and used diesel vehicles will be required to use exhaust cleanup devices similar to those on cars, which will reduce their pollution by up to 98 percent.

Principally, these engineers and scientists work on developing new automotive technologies that will help improve our air quality. They work with **information technology specialists** or have specialized computer-programming skills themselves. They also work with researchers, such as health scientists, epidemiologists, and atmospheric scientists, for example, to better understand

TRANSPORTATION PLANNER

AT A GLANCE

Employment:

11,000

Demand:

Good

Breakdown:

Public sector, 70 percent

Private sector, 30 percent

Trends:

• More experienced transportation planners are taking early retirement incentives, opening up entry-level and middle-level positions throughout the country.

• Public demand is calling for more transportation alternatives, such as pedestrian and bicycle pathways, creating a demand for well-educated, creative planners.

Salary:

Entry-level salaries begin around $33,000, with median annual salaries reaching from $41,000 to nearly $65,000. The top 10 percent averages around $79,000.

JOB DESCRIPTION

Transportation planners facilitate daily transit activities by designing roadway and mass transit plans. They develop, write, and implement transportation policy; interpret local codes and regulations; research the effectiveness, use, and convenience of current transportation plans; and develop public outreach campaigns to determine community transportation needs. "We figure out what things affect travel behavior and work around that," says Brian Gregor, senior transportation analyst and planner for the Oregon Department of Transportation. "We then work with the public and decision makers, like elected officials, to lay out design plans."

Transportation planners design systems that incorporate travel demand, commuter safety, and the efficient use of transport options available to a community. These include cars, bicycles, pedestrian walkways, trains, buses, and even boats. Steven Gerber, a transportation planner for the city of Portland, Oregon, says that his greatest challenge is integrating transportation needs while at the same time encouraging mass transit use for the environment's sake.

As concerns about traffic congestion and pollution grow, many transportation plans feature enhanced bicycle and pedestrian accessibility and encourage public transportation use. These alternatives curb public use of gasoline, as well as pollution generated from single-passenger vehicles. According to state-based Gregor, "People are seeking alternatives to mass transportation, and one way of going about it is to look to bike and pedestrian paths as a way of improving quality of life."

Transportation planners work for local, state, and federal government transportation departments, or sometimes for consulting firms and nonprofit organizations. No matter where they work, these planners meet and collaborate with land use planners, engineers, elected officials, business associations, and citizens to design and implement transportation system plans. They often use technology like data spreadsheets, PowerPoint, geographic information systems, ArcView, and EMME2, a modeling program that allows planners to see and analyze interactions between land use and transportation needs.

GETTING STARTED

• A bachelor's degree in civil engineering or urban or regional planning is the minimum requirement for employment.
• Senior positions require master's degrees in areas such as transportation planning, civil engineering, and urban planning.
• Internships or work opportunities at consulting firms or government offices provide valuable experience working with computer systems, geographic information systems, and other technology used by transportation planners.

ADVICE FROM THE PROS

David Lee, administrator, statewide planner, and policy analyst for Florida's Department of Transportation, suggests finding a niche in the field, such as aviation, mass transit, or pedestrian and bicycle path development. "Find what you're interested in, and focus on that field," he says. "There are more job opportunities for more transportation planners who specialize in something. It's also a good way to get exposure from the ground up."

the effects of air pollution. These researchers evaluate the basic science that supports technological innovation, such as predicting reduced asthma rates or other health benefits that new technology will bring.

Creating emissions requirements and new technology that will achieve such requirements for all products will help improve our air quality in numerous ways. Besides cars, for example, researchers are developing technology and regulations aimed at reformulating consumer products like paints, which contain organic compounds that cause smog. However, improving air quality by way of technological innovation will be effective only if these clean technologies both enhance performance and are offered at competitive prices. For example, consumers have not shown an overwhelming desire to choose between clean or dirty cars based purely on environmental improvement.

Nonetheless, consumers are buying and using less polluting products as they become more available, which has led to relatively rapid emissions reductions in the state of California. This explains why air pollution here has improved by about 50 percent over the past twenty years, while economic and population growth over the same time has been 75 and 40 percent, respectively.

- Generating Incentives

Economic incentives can also help reduce emissions to improve air quality. Specifically, we can set up tax structures that encourage behavior change. Theoretically, by creating higher taxes or prices for pollutant-laden commodities, people will buy fewer of them, or choose less polluting versions, in order to save money. Raising the price of gasoline is one way to discourage driving while encouraging the use of alternative transportation modes and decreasing overall emissions. Another way to discourage driving is by way of better community design, as implemented by **land use planners** and **transportation planners.**

Unfortunately, tax strategies are not usually favored politically in the United States. In contrast, many countries in Europe are successfully using tax policies to broadly encourage alternative lifestyle choices that result in improved air quality. Great Britain, for instance, taxes people's cars based on global warming emissions per car. Similarly, Germany taxes diesel fuel less than gasoline to encourage carbon dioxide emission reductions. These successes prove that environmental taxes can be effective without hurting our economy. However, we have a lot of work to do domestically to implement similar incentives. (For more information about environmental tax reform and economic incentives, see chapter 7 on economics.)

- Quantifying the Benefits

Too often clean air is thought to be too expensive, and legislation aimed at achieving it ends up on the cutting room floor. Economists and finance professionals assess the costs, benefits, and feasibility of introducing clean technology, tax incentives, and government regulation before they are actually implemented. In assessing air pollution's impacts, careful analysis usually shows that the cost of reducing emissions (whether by implementing government regulation, economic incentives, or technological innovation) is much less than the benefits of eliminating those pollution sources, typically by a factor of five or more. Economic and health benefits of reducing air pollution include people living longer, avoiding hospital stays, saving health care costs, and working instead of staying home due to respiratory problems—all of which have value in the billions of dollars and inevitably lead to productivity and public health improvements in our overall society. Implementing government, technology, and economic strategies are ways for people, the natural environment, and our economy to all benefit from reduced emissions and improved air quality. The ultimate result is when air quality everywhere is safe for everyone.

Career Advice from the Expert

Despite all of these improvements, one significant challenge remains: will we eradicate air pollution before human and ecological health is permanently damaged? Aspiring professionals working on strategies already mentioned can help tackle this challenge.

Professionals in this field usually have engineering or science degrees in chemistry, physics, or biology. Some specialists have degrees in economics, atmospheric science, or public policy. For many jobs, especially in government or working to influence government policy, advanced master's degrees are often a prerequisite.

Career Spotlight

Division of Air Pollution Control
State of Ohio Environmental
Protection Agency
Columbus, Ohio

The job of protecting air quality in the United States falls overwhelmingly on state and local governments. Although the Clean Air Act is a federal law, nearly every state has accepted the delegation of authority to deal with emissions from cars and trucks, factories and farms, and businesses and government facilities. In Ohio, the task falls largely on the fifty-two employees of the Division of Air Pollution Control (DAPC) at the Ohio Environmental Protection Agency. From the urban regions of Cleveland and Cincinnati to the hundreds of small towns and rural areas that cover the state, it's their job to ensure that Ohio's 11.5 million people are breathing clean, healthy air.

The heart of the division's approach to air pollution from stationary sources is the permitting section, where Patty Hemmelgarn works as an environmental specialist. Her team receives 1,100 requests annually for permits to approve new air pollution sources. Upon receiving an application, she determines which regulations apply, calculates emissions, and

establishes limits, terms, and conditions for monitoring, record-keeping, and required reporting. These are written into the approved permits, and on-site emissions testing is then used to demonstrate compliance.

"For me, the most interesting days on the job are when I go out to the facilities for emissions tests or inspections," Patty says. "Face-to-face meetings with the facility's environmental supervisors and emissions unit operators help to create a partnership for clean air." The meetings also greatly improve compliance because violations are often discovered during inspections. When this happens, she works with facility staff to resolve problems by working together. She refers cases for enforcement action when necessary.

Patty's education was good preparation for the job. "Because I understand how most emissions units operate, I can calculate emissions, craft appropriate operating restrictions, and evaluate whether an emissions unit is operating correctly," she reports. "I like working with each piece of the air compliance puzzle—from permits to emissions tests to compliance inspections, and my B.S. degree in chemical engineering from Ohio University was great training."

Reducing air pollution from cars, trucks, and other vehicles is the job of the mobile sources section. "Vehicle exhaust contains carbon monoxide, nitrogen oxides, volatile organic compounds, and other pollutants that harm people and wildlife," says Steve McVey, one of the section's environmental specialists. "And there are a lot of vehicles on Ohio's roads."

He's right. The section's motor vehicle inspection program tests two million vehicles each year, measuring tailpipe exhaust composition, using on-board diagnostics, and collecting inspection station data.

After graduating from Ohio University with a B.S. in chemical engineering, Steve worked as an operations engineer for companies producing industrial chemicals. Seeking a change, he obtained a master's degree in chemical engineering with an emphasis in computer modeling. "My background in computers and engineering comes into play as I record, manipulate, and analyze data. The academic experience has been a really valuable asset."

Today, his job includes website development and computer modeling of mobile source emissions. Together with state transportation planners and local agencies, he sets parameters such as climatology data, vehicle miles traveled, and vehicle age distribution, and simulates emissions under various scenarios and control strategies. Ultimately, these results help state legislators decide which pollution control programs to adopt.

Debbie Bradley's path to her job as a quality assurance auditor in the air monitoring unit took a few turns along the way. While majoring in biology at Bowling Green State University, she met an Ohio EPA representative at career day. Interested in pursuing environmental protection, she changed her major, graduating with a B.S. in environmental science. At first, she accepted a position conducting water analysis in a laboratory setting. However, she was unhappy working in a building with no windows. "I decided to get my foot in the door with Ohio EPA by volunteering," she recalls. "That summer, I gained experience outdoors, mapping and determining the health of rivers and streams, and realized I wanted a job in the field." Hired by DAPC nine months later for an air monitoring position, she is now on the front lines of Ohio's air pollution control efforts—ensuring that the information that drives all environmental decisions and planning is of high quality.

Because professionals use scientific measurements, economic data, and computer modeling to establish effective policies that are often based on solid statistics and other data trends, those looking to work on air quality issues will benefit from technical experiences like participating in lab measurement or air quality monitoring. Look for opportunities to volunteer, intern, or work for universities that have research projects using air quality or emissions equipment, or with government agencies that expose prospective professionals to measurement, data collection, and analysis. Sometimes this technical experience can also lend a firsthand look at how information is used to formulate policy. In any case, employers value practical experience, and even just a little can give you an advantage.

As with many professions, excellent communication skills are essential for this work. Air quality workers must be able to convey their findings, advocate for specific changes, and explain why those changes are necessary. Communication skills are especially important when presenting ideas to people who do not have technical expertise. Most professionals are also involved in creating policies that determine how air pollution will be cleaned up, which typically requires extensive interaction between government regulators, businesses, and public interest groups.

One of the best ways to develop good communications skills is by volunteering for a public interest group involved in air quality issues. Here, you may have the opportunity to write issue papers, prepare speeches, or draft articles for newsletters or brochures. Many technical people do not usually consider working in this arena because it is not always science-driven. However, volunteering for such organizations or interning with local congressional representatives will give aspiring air quality professionals an opportunity to understand how policy affects business and government, and ultimately, people.

RESOURCES

Airhead, www.airhead.org

American Indoor Air Quality Council, www.indoor-air-quality.org

American Lung Association, www.lungusa.org/air

California Air Resources Board, www.arb.ca.gov

Center for Clean Air Policy, www.ccap.org

Clean Air Trust, www.cleanairtrust.org

Environmental Protection Agency Office of Air Quality Planning and Standards, www.epa.gov/oar/oaqps

Ohio Environmental Protection Agency, www.epa.state.oh.us

Union of Concerned Scientists, www.ucsusa.org

South Coast Air Quality Monitoring District, www.aqmd.gov

State and Territorial Air Pollution Administration, www.cleanairworld.org

3. ARCHITECTURE, CONSTRUCTION, AND DESIGN

Architect William McDonough.

A CONVERSATION WITH
William McDonough

Nature can be inspiring. Take it from William McDonough, founder and principal of William McDonough + Partners, one of the nation's leading architecture and community design firms. "Nature is a source of both sustenance and exquisite design," he says. "When architects follow the laws of nature, their designs can celebrate rather than lament the human impact on the natural world."

McDonough is best known for developing a principled design approach that integrates diverse economic, social, and cultural criteria within an ecologically intelligent framework. In 1999, *Time* Magazine recognized him as a "Hero for the Planet," because "his utopianism is grounded in a unified philosophy that—in demonstrable and practical ways—is changing the design of the world."

McDonough obtained an M.A. in architecture from Yale University. He has been featured in numerous articles and books and writes frequently on architecture, design, and sustainability. Most recently, with the German chemist Michael Braungart, he co-authored *Cradle to Cradle: Remaking the Way We Make Things* (North Point Press, 2002).

What Is the Issue?

Many of the environmental problems of mainstream architecture, construction, and design are the result of a divorce between human designs and the natural world. Modern design evolved during the first industrial revolution with the advent of fossil fuels, concrete, and large sheet glass. Rather suddenly, buildings became sleek, mass-produced machines. **Architects** forgot the sun, they forgot the surrounding world, they forgot the natural flows of water and energy. Consequently, current buildings are one size fits all, with artificial climates and little relation to place. Notwithstanding various stylistic exercises, buildings are the same in Europe, Malaysia, and Houston, which is problematic. We are simply adding fossil fuels to ameliorate climate conditions: in hot and humid places we air-condition, and in cold, dry places we heat. Obviously people seek comfort, but conventional buildings provide that comfort by disengaging from their climate and surrounding environment—often with very unsatisfying results.

- Hazardous Materials

There are also other problems. Very few materials used for building interiors are specifically designed for indoor use, and, consequently, indoor air quality in conventional buildings is often poor. Many paints, textiles, adhesives, and carpets give off gases that are volatile organic compounds, some of which are suspected carcinogens and immune system disrupters. Other materials are produced with toxic or questionable ingredients, such as PVC, which contains plasticizers suspected of disrupting human endocrine systems, and heavy metals that are known to be carcinogenic. Due to some materials' toxicity, their disposal is also potentially hazardous. For more information about toxics, please see chapter 19 about toxics and human health.

- Environmental Impacts

In our current architecture, design, and construction system, buildings also have immense impacts on the larger global environment. These impacts are visible in various forms, perhaps most easily in the sourcing of materials. For example, every ton of copper mined produces 400 to 600 tons of waste, resulting in natural resources and healthy landscape losses. Careless wood harvesting can cause forest destruction. Even forms of genocide can be attributed to extractive industries, as when entire tribal cultures are wiped out by the harvesting of virgin mahogany forests in which they live. Existing conventional buildings also consume an inordinate amount of energy and water, and account for one-half of the world's waste stream. New construction consumes some three billion tons of raw materials each year.

- Poor Community Design

The design problems evident in single buildings are repeated in the design of communities. Current land use and planning models rarely see communities in the overarching context of the natural world, and so, new development often creates habitat loss, pollution, and other forms of environmental degradation. At the same time, the rubber-stamping of conventional sprawl fails to take into account the social and economic dimensions of communities. The separation of residential and commercial neighborhoods is one notable example. Diversity is another casualty. We need many kinds of communities to serve different people's needs. Why not design rich, mixed, livable communities that create intense, positive interactions with the environment rather than one-size-fits-all developments divorced from the natural world?

Problems in planning also mean problems in our cities. Urban areas are the world's population centers. Designing livable, productive, delightful cities is an essential part of our planet's future. Currently, most cities are not ecologically intelligent or optimally designed. They don't benefit from interactions with the natural world—they rarely tap into unlimited solar energy, for example.

- Regulation as Design Failure

Regulation is often seen as an answer to our environmental woes, but it does not solve environmental problems. Instead, regulation is a signal of design failure, a form of negative feedback from society indicating that certain materials and practices are not optimally designed. Certainly, regulations and codes of conduct are essential to the practice of architecture. Whether they regulate building performance, prescribe specific stylistic desires of a community, or voluntarily answer public outcries for environmental protection, codes institutionalize a wide spectrum of social feedback on the impacts of architecture. However, when industries codify regulations, innovation often takes a back seat. Rather than being a generator of new designs that seek to integrate social and economic concerns into ecologically intelligent frameworks, "meeting code" is often an exercise in meeting minimum expectations. Rarely does it lead to the optimal use of energy and materials.

The U.S. Green Building Council's Leadership in Energy and Environmental Design (LEED) standards show that industry professionals are interested in addressing environmental concerns. But while the LEED standards have successfully created considerable environmental improvements in architecture, design, and construction, they are still a consensus-based exercise, subject to constant review, testing, and watering down by larger group input. As such, they can still end up supporting a "less bad" approach to architecture rather than generating creative responses or bold innovations in the use of energy and materials, both of which are greatly needed in the world of architecture.

Regulations also do not always benefit society. The carpet industry, for example, wants recycled content to be its only regulation metric. But this ignores the quality, content, and potential hazards of the materials themselves. In carpets, much recycled content is PVC, making many recycled carpets potential toxins. If this metric becomes the regulatory standard, we will perpetuate poor design and a dangerous system.

- ## Barriers within the Field

Probably the greatest barrier to change—for better or worse—is the infrastructure of specifications and codes, which are resistant to innovation and can make the whole building process a struggle with inertia. Some of the barriers to change can come from within the field. Today, young professionals are entering the workforce with great skills and high expectations and with more practical experiences than ever before, and they want to be involved in innovative and meaningful work. Many of them understand and take for granted that architecture must be ecologically intelligent. Their professors, however, may have had little practical experience to offer; midcareer professionals who are their superiors are also not always on the same wavelength. For example, the average professional is somewhere from twenty-five to sixty-five years old. Any midcareer engineer, software designer, **architect**, or other professional has had a different education than young people today, yet they are the senior managers who make high-level decisions that sometimes are impediments to

change. To make matters worse, litigious climates make the average professional very risk-aversive and unlikely to try a new system until it has some kid of
performance track record.

How Are Environmental Professionals Approaching the Issue?

While changing conventional practices can seem like a daunting task, buildings are the result of human design and therefore design can also offer solutions. Design is the first signal of human intention. Rather than setting out to
create "less bad" architecture, it makes more sense to aim, right from the start,
to design buildings that support life by generating only positive effects. We can
even design buildings like trees: buildings that make oxygen, sequester carbon,
fix nitrogen, distill water, provide habitat for thousands of species, accrue solar
energy as fuel, build soil, create microclimates, and change with the seasons.
That's quite a change from the inanimate, one-size-fits-all buildings that conventional practice delivers.

• Reimagining Design

Architects and designers have begun to create buildings such as these by reimagining the relationship between human designs and their surroundings. A

BOX 3-1

REINVENTING ARCHITECTURE, CONSTRUCTION, AND DESIGN

- Reimagining Redesign
- Redesigning Building Materials
- Designing for Disassembly and Conversion
- Thinking Innovatively
- Benchmarking
- Making a Team Effort

building, like a tree, fits within its particular locale. It doesn't overwhelm its set-ting; it supports and depends on the local ecosystem. The relationship between building, site, and place is interactive and regenerative. A building we designed for Oberlin College, for example, purifies water with botanical gardens, uses geothermal wells for heating and cooling, is protected from winter winds by native trees, and taps the energy of the sun with 3,700 square feet of photo-voltaic panels. With energy requirements nearly 80 percent lower than those of standard academic buildings, the Oberlin building will one day generate more energy than it consumes.

That's just one example. We are also seeing buildings with living roofs that filter storm water, restore habitat, and provide thermal insulation. High-rises are outfitted with solar collectors. Communities are oriented with streets and buildings so that homes and offices can be lit by the sun and cooled by evening breezes. Corporate campuses have rainwater collection systems that capture millions of gallons of water for landscape irrigation. Cities are investing in rooftop gardens, solar and wind power enterprises, and the restoration of watersheds. We are witnessing the emergence of a new paradigm that sees nat-ural systems as a model for human designs.

• Redesigning Building Materials

Building materials are undergoing a transformation as well. Due to the ques-tionable environmental impacts of many materials, **architects** and interior designers are beginning to take a closer look at the materials they specify. Researchers and **chemists** assess materials for various characteristics, rang-ing from chemical ingredients' health and safety impacts to the environmental impacts of their use, harvesting, or manufacture. **Architects** searching for safe, healthful, environmentally sound materials can be helped by organiza-tions such as the Forest Stewardship Council, which certifies sustainably har-vested wood products, or design firms like McDonough Braungart Design Chemistry (MBDC), which conducts thorough chemical assessments and iden-tifies beneficial product ingredients.

Many suppliers are deeply involved in redesigning materials. **Chemists** at Shaw Industries, the world's largest producer of commercial carpet, have designed a perpetually recyclable, completely healthful carpet tile. Everything that goes into the nylon carpet fiber and polyolefin backing has been assessed

ARCHITECT

AT A GLANCE

Employment:

113,000

Demand:

Good

Breakdown:

Self-employed, 30 percent

Private sector, 60 percent

Public sector, 10 percent

Trends:

• Increasing emphasis on sustainable design and the use of environmentally friendly materials for private and commercial development is creating more demand for "green" architects.

• Drafting technology and computer-aided design skills are essential, and competition for jobs is fierce.

Salary:

Entry-level salaries start at $37,000, with the median incomes ranging from $45,500 to $77,830 annually. The top 10 percent earns more than $95,000.

JOB DESCRIPTION

First and foremost, architects are creative thinkers who assess clients' needs and then conceptualize and design buildings accordingly. Today, architects are turning to "green architecture" practices to address the growing concern about environmentally hazardous materials used in construction, says architect Emily Mensone Turk of the U.S. Green Business Council (www.usgbc.org).

Architects are starting to incorporate building materials made from recycled products into their designs, as well as sustainable site planning, energy and water efficiency, and environmental impact concerns. The challenge lies in "recognizing that sustainable design must be part of all facets of life to be successful," adds Turk. "I'm interested in energy and resource allocation as well as good air quality and human well-being."

Architects involved in predesign activities conduct feasibility and environmental impact assessments, as required by local cities and towns. After consulting with clients, they draft plans showing a building's outer appearance and construction details, including plumbing and electrical systems and landscape plans. These plans must follow building codes, fire ordinances, zoning laws, and other municipal regulations. During this process, many architects use computer-aided design systems rather than traditional pencil and paper sketches to complete their designs.

James M. Hanifan often specializes in preserving environmentally sensitive wetlands and wildlife as an architect for Caolo & Bieniek Associates, Inc., a firm based in Chicopee, Massachusetts, that focuses on building schools and other public facilities in New England. "A common factor in cities and towns is that most private developers buy up the best parcels of land in the area," he says, "so cities and towns wanting to build public facilities are often left with land that's more difficult and costly to build. As a result, we must keep a balance of natural land and development."

GETTING STARTED

• A bachelor's or master's degree in architecture is required for employment. Most states require that professionals obtain degrees from schools accredited by the National Architectural Accrediting Board (www.naab.org).

• Architects must pass the Architect Registration Examination (ARE) to become licensed professionals, so it is important to seek internships that will help you prepare for this exam.

• The U.S. Green Building Council has developed the Leadership in Energy and Environmental Design (LEED) standards to encourage sustainable architecture, design, and construction. Participating in this program will give you an advantage for finding employment, since LEED is highly regarded as an ethical and high-performance marker.

ADVICE FROM THE PROS

Architect James M. Hanifan says that students interested in architecture must be creative thinkers, ambitious, patient, and detail-oriented. "As you're moving through the design of a product, you must pay attention to details and follow the project all the way through to the end."

for health and safety. The high-quality ingredients of both materials are designed to be chemically recycled, raw material to raw material, and used again and again in high-quality carpets, rather than being down-cycled into products of lesser value.

This is revolutionary. It allows us to redesign high-tech architectural materials so they can be used safely and effectively in every phase of building construction—and again in new buildings. Geopolymers, for example, are a promising replacement for concrete, which leaches harmful chemicals on buildings sites and in landfills. Made from local earth and high-quality plastic, geopolymers are far more stable than concrete and require far less energy to produce. Design for disassembly allows building materials made of geopolymers to be used again in new buildings. Materials scientists at BASF are developing another high-tech material designed for reuse that is made from polystyrene foam, a structural building material for low-cost housing in developing countries.

- ## Designing for Disassembly and Conversion

When we design materials for disassembly we can purposefully design buildings for future deconstruction. A materials passport—a tracking code created with molecular markers, for example—can guide materials through industrial cycles, routing them from production through reuse, defining optimum uses and intelligent practices. With a materials passport we know how to take products apart, where valuable and recyclable materials are, and how these materials could be reused and reassembled. Valuable construction materials would be used *again* as valuable construction materials, not recycled into hybrid materials of lesser value heading inexorably toward the landfill.

- ## Thinking Innovatively

Chemical recycling and materials passports suggest the large role technology plays in the design of ecologically intelligent architecture. Using nature as a model for human design does not mean shunning technology; it means creating effective, life-affirming building and energy systems.

Consider Ford Motor Company's new manufacturing facility at its famous Rouge River site in Dearborn, Michigan. The new River Rouge plant has a green roof—a ten-acre surface covered with a thin layer of soil planted with sedum, a noninvasive succulent that does not require irrigation and protects the building from thermal shock and ultraviolet degradation. This is no small thing. Whereas the temperature on conventional roofs can fluctuate between 70 and 140 degrees Fahrenheit, decreasing their durability, green roofs moderate temperature on the rooftop and in the building itself and can last indefinitely.

The Rouge's green roof also adds value by creating an effective storm water management system. The soil and sedum absorb storm water, provide free evaporative cooling, and channel rainwater into porous parking lots that filter runoff into surrounding constructed wetlands. The entire landscape is a giant biofilter. It's also cost-effective: the entire system cost $10 million less than the conventional technical controls that would have been needed to manage storm water runoff at the site.

The Ford plant maximizes natural air circulation by employing ten rooftop heating, ventilating, and air-conditioning (HVAC), units whereas conventional designs would have employed thirty-seven small units. The Ford plant uses only ten because the entire building is designed as a duct. It delivers much higher air quality and temperature control because the building is positively pressurized to augment temperature changes caused by seasonal temperature differences when truck dock doors are open.

At the worker level, the River Rouge does not churn out bad air by channeling it through oil mist from drilled ceiling holes. Instead, its air-conditioning system consists of a simple bubble on the roof that allows air to drop naturally throughout the building under computer control, giving workers ample daylight and healthy air quality, which typical buildings rarely provide. At McDonough + Partners, we would rather design a life-support system for people who work than a work-support system for people who don't have a life.

- Benchmarking

Projects like these can be used as leading examples in reimagining architecture, construction, and design. Government can look for examples of cost-effective buildings that benefit society, the environment, and the economy, and then put these models forward for other people to emulate. This type of benchmarking will not only create recognition and praise for cutting-edge projects, but also allow government to play a positive role in creating new, environmentally effective ideas, rather than creating regulation for "less bad" projects.

As more of these concepts mature and more buildings are built, it will be easier for people to access the creativity and innovation of ecologically intelligent design. For example, by bringing green roofs into mainstream design, we are removing the liabilities for implementing such ideas. Five years ago, if a young **architect** walked into a firm and said, "I think we can build a green roof," he or she would have met considerable resistance because there weren't any in the country. Now, however, thousands of **architects** have seen successful examples that have met budget and time restrictions. When an **architect** suggests a green roof today, his or her superiors have heard of this idea and know that some of the smartest firms are creating them. By completing more projects, we can lead by example.

The most powerful incentive for implementing alternative buildings is financial. Because innovative projects have been immensely profitable, we are creating the ultimate incentive for any business. So cost-effectiveness must be integrated into a design as much as eco-effectiveness. As more projects are created, project financial consultants can demonstrate to clients how they can save money and generate lasting value, just as Ford has done.

• Making a Team Effort

Effective design team is key to bringing new, innovative projects to fruition. Architects are typically generalists. When someone considers a career in architecture, the environmental component is going to be one small part of his or her training, which is a good thing, because even environmentally oriented **architects** must be trained in the overall discipline. As generalists, **architects** act as design team leaders in order to connect and incorporate culture, technology, the arts, science, and community into any design. Other design team participants are specialty professionals including **landscape architects**, who understand and use ecology as a tool to integrate buildings into landscapes, as well as botanists, **chemists**, day lighting specialists, and energy specialists, including solar and mechanical energy professionals. Teams of contractors and engineers who understand ecologically intelligent design and the potential for new technology are working with these professionals. Additionally, design teams now comprise a very rich mix of consultants. It's an exciting, promising time to be entering the field.

LANDSCAPE ARCHITECT

AT A GLANCE

Employment:

23,500

Demand:

Excellent

Breakdown:

Public sector, 20 percent

Private sector, 80 percent

Trends:

• Growth in residential, commercial, and public works construction is creating employment opportunities.

• Increasing public demand for restoration and preservation of natural areas disturbed by development is fueling job growth.

Salary:

Starting salaries begin around $30,000, and median earnings equal $49,330. Most landscape architects earn between $35,715 and $64,000. Senior people earn $82,000 or more.

JOB DESCRIPTION

Restoring landscapes torn apart by urban sprawl. Healing ecological damage in the Florida Everglades. Repairing forests devastated by fires ripping through the American West. These are among the many environmental challenges creating jobs for landscape architects (L.A.s). "Our field attracts creative people working on commercial and residential design, urban and regional planning, and environmental protection and restoration," says Ron Leighton, director of education and academic affairs at the American Society of Landscape Architects (ASLA). Traditionally known for landscape design at private homes and businesses, or for constructing elaborate gardens, today's L.A.s are likely to serve as active stewards for public landscapes as small as city parks and as large as entire watersheds or ecosystems. Landscape architects draft detailed site plans and budgets, prepare land use studies and plans, create models and cost estimates, and manage laborers and technicians on project sites. To tackle this wide variety of tasks, an L.A. needs keen awareness of how natural systems work, including solid knowledge about soil composition, hydrology, climate variations, plant physiology, use of native plants, and ecological factors. Technological advances have revolutionized landscape architecture and created a new set of required skills. "Knowledge of technology like geographic information systems (GIS) such as ArcView, Adobe Photoshop, and computer-aided design programs like AutoCAD is essential for landscape professionals," stresses Leighton. Scientific and technological knowledge, however, is not enough. L.A.s must also collaborate with clients, local officials, architects, planners, engineers, and others; being "good with people" is essential for success. A majority of landscape architects work in private firms, while a small but growing number hold jobs at local, state, and federal government agencies.

GETTING STARTED

• A bachelor's degree in landscape architecture is required for employment. Programs accredited by the Landscape Architecture Accreditation Board of the American Society of Landscape Architects (www.asla.org) are highly regarded.

• A master's degree in landscape architecture is highly recommended, especially for professionals aspiring to own their own firm.

• Forty-six states require landscape architects to be licensed or registered. Licensing is based upon the successful completion of the Landscape Architect Registration Exam (L.A.R.E), usually requiring a degree from an accredited school and one to four years of experience working with a licensed landscape architect. Specific licensing standards for each state may vary slightly. This information is available from the Council of Landscape Architectural Registration Boards (www.clarb.org).

ADVICE FROM THE PROS

Mark Papa, a landscape architect who owns a small firm in Connecticut, encourages professionals who are new to the field to "switch firms frequently before settling down. Small and large firms take on different types of projects and this approach exposes you to the type of work that you may want to undertake in the future. Multi-project exposure results in flexible, well-rounded landscape architects—the best kind!"

Career Spotlight

William McDonough + Partners
Woods Hole Research Center
Green Building Project
Falmouth, Massachusetts

Creating buildings that actually deserve to be called "sustainable" requires a partnership among many different companies and professions—architects, planners, landscapers, contractors, engineers, suppliers, and more. The design and construction of the Woods Hole Research Center's (WHRC) new facilities in Falmouth, Massachusetts, is a perfect case study of what goes into the development of environmentally sound architecture in the twenty-first century. Among other challenges, the project involved the conversion of a nineteenth-century summer home on Cape Cod into a state-of-the-art research facility.

As an organization dedicated to studying the effects of human activities on the environment,

The conference room of the Woods Hole Research Center embodies sustainability principles like maximum use of natural light, while creating an atmosphere of beauty and elegance. Source: Judith Watts Photography.

the WHRC wanted to demonstrate how modern building construction can "harmonize with a habitable earth." For example, the center is committed to reversing global warming and wanted its building to use renewable sources of energy, or even produce more energy than it consumed on a net annual basis. Because William McDonough + Partners (WM+P) shared the center's vision of creative solutions, the WHRC selected WM+P to lead the design.

The firm turned the management of the job over to associate partner Mark Rylander. Rylander has made a career out of managing the design and construction of complex, environmentally intelligent building projects, bringing diverse perspectives together to create architecture that "walks the talk." For Woods Hole, Rylander provided project management, team leadership, and continuity. Rylander received a B.S. in architecture form the University of Virginia and an M.A. in architecture from Yale University. Currently the 2004 chair of the American Institute of Architects, Committee on Environment, he focuses on reestablishing positive relationships between human communities and the natural world. Rylander remembers, "I worked closely with [WHRC] to create a building that is not only incredibly energy efficient but also offers abundant daylight, fresh air, and views of the forest."

To address landscaping questions, Rylander selected Warren Byrd, principal at Nelson Byrd Landscape Architects and a graduate of Virginia Tech (B.S./horticulture) and the University of Virginia (Masters of Landscape Architecture). Byrd's firm collaborates frequently with nationally recognized green architects on a diverse portfolio of projects. "I emphasize the importance of understanding the local characteristics of the physical and geographic regions within which we work," Byrd says. "Natural systems and plant communities offer a unique base of knowledge that influence our design concepts."

Working closely with a civil engineer, Byrd led the development of the center's ingenious landscape master plan, which simultaneously preserves habitat, manages storm water onsite, and integrates the building within the coastal ecology of its setting.

The design for the center's new campus also needed to give shape to its vision for the future in a high-performance building that embodies exceptional environmental intelligence. To achieve this goal, Mark Rylander engaged Marc Rosenbaum, an energy analyst. "Rosenbaum was invaluable in instructing us on the building envelope," Rylander remembers. "He showed us how to create a building that was incredibly energy efficient without being a drab, windowless, insulated box."

A professional engineer and a nationally recognized pioneer in sustainable energy and energy efficiency, Rosenbaum uses an integrated systems design approach to create buildings and communities that connect people to the natural world, supporting personal and planetary health. Realizing that the barriers to high-performance buildings and communities are neither technical nor economic, his work includes helping clients to design the processes necessary to create high-performance projects. His ability to communicate energy and engineering concepts in clear commonsense language has been essential to his success. Educated with B.S. and M.S. degrees in mechanical engineering from MIT, he is a licensed engineer and a U.S Green Building Council/LEED accredited professional.

The balance of the WHRC building team also included an array of nationally known professionals including, among others, a structural engineer—Nat Oppenheimer of Robert Stilman Associates—with celebrated expertise in historic structures; lighting designer David Nelson of Clanton Associates, who brought experience in energy efficient lighting; and T. R. White, a young construction firm from Boston whose excellent work on the WHRC has led to a new series of green construction projects.

Completed in 2003, the center's new home makes use of a range of passive and active strategies. It uses natural lighting to optimize aesthetics and energy demand, high-efficiency building systems, fixtures, and equipment as well as renewable geothermal energy to heat and cool the building. Like the forests the center studies, the building draws its power from energy harvested from sun and wind, via a twenty-six-kilowatt photovoltaic array and a proposed on-site wind turbine. Appropriately, the building itself now serves as an educational model, encouraging thoughtful approaches toward materials selection, energy efficiency, water management, and site design capable of being replicated by others in the region.

Career Advice from the Expert

Those interested in the environmental side of these professions must be trained in conventional practices. Once they obtain basic skills, they should look for innovative technologies and practices in order to integrate these concepts into their designs after developing professional standing.

Young professionals should also become educated in economics and business so they can communicate new ideas in the language of business clients. For example, when speaking with Bill Ford, I never presented the redesign of the River Rouge plant as a moralistic issue about how to reduce its ecological footprint. This would be an example of saying "You should be less bad," and people don't respond to that. Instead, completely innovative ideas urge us to throw away the "less bad" concepts entirely and create different structures. In order for new ideas to appeal to business, they must be profitable, cost saving, and environmentally or eco-effective. The fact that ducks and songbirds are returning to the River Rouge is wonderful and shows the design's ecological intelligence. However, I never presented this benefit to Bill Ford as the reason why he should hire me. Rather, I suggested how a project like this saves money, creates new investments, and improves environmental conditions.

There is no particular route that someone entering this field must follow, although conventional design training is a necessary precursor. Having a deep understanding of society and its relation to the environment is critical for enjoying this work. As Joseph Campbell says, "Follow your bliss." Young professionals should not think they are going to "save the world," but rather decide what they want to do and where they want to work. They can then celebrate what they are doing, be grounded by their professional training, and bring that knowledge to the world.

RESOURCES

American Institute of Architects, www.aia.org
American Society of Landscape Architects, www.asla.org
EnivironDesign, www.isdesignet.com/ED
Green Blue Institute, www.greenblue.org
Green@Work Magazine, www.greenatworkmag.com
Interiors and Sources, www.isdesignet.com

McDonough Braugnart Design Chemistry, www.mbdc.com
Metropolis Magazine, www.metropolismag.com
National Building Museum, www.nbm.org
U.S. Green Building Council, www.usgbc.org
William McDonough + Partners, www.mcdonoughpartners.com

4. BIODIVERSITY LOSS AND SPECIES EXTINCTION

Stuart Pimm, off to survey endangered species in Everglades National Park, Florida.

A CONVERSATION WITH
Stuart Pimm

Years of exposure to the natural world's relentless destruction have left Stuart Pimm feeling . . . hopeful. "There's no need to throw yourself off of a building," he says. "News about Earth's losses can be terribly depressing, yet I'm unashamedly optimistic. The planet is not healthy, but it's not fatally wounded. We are not doomed."

Pimm's career defending nature is uniquely wide-ranging. A respected research conservation biologist and university professor, he also works creatively to merge science, practice, and policy. As if that weren't enough, he has published over 150 articles and produced engaging speeches and books, including *The World According to Pimm: A Scientist Audits the Earth* (McGraw-Hill, 2001).

Pimm obtained a Ph.D. from New Mexico State University and recently taught at Columbia University. Currently, he is the Doris Duke Chair of Conservation Ecology at Duke University, and he maintains active field research in the Everglades, Madagascar, Brazil, Central America, and South Africa.

What Is the Issue?

There are over six billion people on Earth today. We are causing massive damage to Earth's ecosystems and the planet's capacity to support its extraordinary biodiversity. We clearly are driving some species to extinction before their time. This is happening at least at 1,000 times the natural or geologic rate, and although some people express doubts, these elevated rates have been documented and cross-checked by scientists many times. In fact, the real rate of extinctions may be much higher. Humans are responsible for species extinction rates that are higher than any time since the dinosaurs disappeared, some 65 million years ago.

Moreover, some species are disappearing before we even know they exist. We don't know the names of 90 percent of the ten million species of plants, animals, birds, fish, and insects on the planet. When scientists and volunteers seriously go looking for species in remote or even familiar places, we find new species that we knew nothing about.

Extinction, of course, is only one way to measure biodiversity threats. A group of scientists from Bird Life International in Cambridge, England, has cataloged birds throughout the world that are either known to be extinct, presumed to be extinct, critical, endangered, or vulnerable. In 2001, this catalog contained 1,100 species of birds. Here in North America, birds are not so badly off—only some 6 percent feel immediate threats. But other species are in serious trouble. The Nature Conservancy tells us that 43 percent of this continent's freshwater mussels, 37 percent of crayfish, 23 percent of amphibians, and 21 percent of freshwater fish have gone extinct or are in immediate danger.

We are appropriating an incredible percentage of the natural world for human use. By some estimates, the world's six billion people use or degrade up to 40 percent of all plant growth, 50 percent of freshwater, and one-third of life in the ocean. By the middle of the century, there will be up to ten billion people, most seeking a standard of living that Americans and Western Europeans enjoy. We must act now if we are serious about protecting biodiversity.

• Causes of Extinction

In the business of preventing extinction, it's important to understand the causes. The most important include:

1. Habitat destruction
2. Pollution

3. Overharvesting or overhunting
4. Habitat fragmentation
5. Introduced species

Some of these damages are clear and obvious, like degrading habitats by clear-cutting a forest, damming a river, poisoning the water and soil, or destroying whole forest and prairie ecosystems and replacing them with crop-lands. Large-scale habitat destruction of these kinds led to the extinction of forty-three known native birds and eighty-four native plant species in Hawaii, for instance. Additionally, global climate change can now be added to the list.

Hunting a species to death is also an obvious problem. In Africa, poaching has contributed to endangering the black rhino population over the past century. People hunt rhinos for various reasons and social pressures. For example, in the Far East, demand exists for powdered rhino horn, which is an ingredient of traditional medicines to reduce fever. In some Arab nations, curved daggers with rhino horn handles serve as status symbols. Poor people in African countries, even those charged with protecting wildlife, have found rhinos a tempting target as well. Civil unrest, corruption, and other societal factors exacerbate these situations. As a result, rhino specialist Holly Dublin and conservation consultant Alison Wilson report that the black rhino population is now about 2,600 animals, an overwhelming decline since the early 1900s when African savannas may have held as many as one million.

Other causes of species extinction are subtler. The role of habitat fragmentation in threatening species is in that category. It may appear that a forest, for instance, has only been partially damaged, but even small changes can deeply affect wildlife's ability to find food, raise their young, migrate to breeding grounds, and stay safe from predators.

People moving alien species all over the planet has often created negative impacts for local plants and animals. Introduced herbivores or carnivores can drive out those already there. Species may behave genetically differently in a new environment, often eliminating an existing species and creating new hybrids—a dangerous effect, as natural ecosystems may not be able to adapt to such changes.

Finally, humans make financial trade-offs in everyday interactions, contributing to biodiversity loss. For example, cutting down a forest to build homes causes property values to increase, even though the ecosystem consequences of this action can be great. As forests are cleared, nearby streams will fill with

Table 4-1. Pressures on Biodiversity

Sector	DIRECT Positive	DIRECT Negative	INDIRECT Positive	INDIRECT Negative
Agriculture and plantation forests	• Creation of diverse ecosystems • Support for biological functions	• Natural ecosystem conversion to agriculture or forest • Fragmenting habitats • Introduction of non-native species	• Maintenance of ecosystem services, enrichment of biological diversity in some cases through mono-culture use	• Pollution of ecosystems through farm chemical runoff • Genetic homogenization • Erosion, siltation, etc.
Fisheries		• Destruction of habitats through damaging fishing practices • Potential overfishing of target species or bycatch species • Introduction of nonnative species		• Pollution of marine and freshwater ecosystems through effluent discharge, excessive nutrient and chemical loading (aquaculture), noise, etc.
Forestry		• Habitat loss or fragmentation through forest clearing and infrastructure construction		• Pollution of forest ecosystems through effluents and noise • Erosion and associated effects • Colonization of natural areas facilitated through infrastructure/access provision
Oil production		• Pollution of ecosystems through spills • Destruction of eco-systems through infrastructure	• Decreased dependency on renewable natural resources (e.g., wood fuel	• Pollution of ecosystem through extraction (e.g., effluents, noise, etc.)
Mining		• Pollution through leaching, etc. • Habitat destruction through infrastructure construction	• Reduced resource extraction through recycling	• Pollution of ecosystem linked to use of inputs in extraction (e.g., effluents, noise)
Transport and related infrastructure		• Facilitates access to fragile ecosystems, fragments habitats, pollution, etc. • Use of land for trans-port infrastructure	• Brings people to conservation sites, increases awareness	• Pollution associated with transport use, including greenhouse gas and air pollution emissions
Water and sanitation	• Creation of special habitats	• Water pollution and overuse destroys habitats and ecosystems	• Water conservation measures beneficial to ecosystems	
Industry		• Pollution of ecosystems • Loss of habitat through infrastructure development		

Source: OECD, 2002

sediment and lose fish. Flowing to the sea, sediment may smother an offshore coral reef, its fish, and beautiful corals; and then obscure invertebrates containing the cure for cancer; or it may threaten reefs that protect inshore areas against storms. Additionally, the forest is a carbon sink. Its loss may accelerate global warming and rising sea levels.

In contrast to destroying a forest to build homes, protecting forests enhances biodiversity by providing ecosystem services such as air and water purification, flood minimization, protection against soil and coastal erosion, crop pollination, climate stabilization, aesthetic pleasure, and many other features.

At a simple level, protecting biodiversity is about slowing or reversing factors related to species extinction. Because the causes of extinction interact with one another in complex ways, reversing biodiversity loss will require equally complex strategies to guarantee we give our children a planet as vibrantly beautiful as the one we inherited.

How Are Environmental Professionals Approaching the Issue?

Protecting biodiversity does not require the same level of urgency everywhere in the world at the same time. It's not exactly synonymous with protecting general ecological health, although perhaps we wouldn't have the current crisis if we valued that more highly. For example, the millions of species on Earth are not distributed equally on the globe; nor are they all equally rare; nor are they all by any means endangered.

The distribution of biodiversity, in particular, is interesting. Probably half of all terrestrial species are located in one-tenth of the world's land area—especially, but certainly not exclusively, in the tropics. Species restricted in high proportion to concentrated areas are *endemics*. Concentrated areas rich with endemic species are called "hot spots." If humanity focused judiciously on these areas, we might save more species. However, governments with such species, like Madagascar and the Philippines, protect less than 2 percent of their land. The same holds regionally in the Algulhas Plain, on the southern tip of Africa and one of the world's "hottest" spots for plants, where an area only half the size of Rhode Island houses 1,751 species. Giving special attention to hot spots can offer extraordinary dividends. Although there are limited resources, many research efforts to identify, map, and explore areas should receive top priority to respond to the speed at which diversity is declining.

As an aside, it's depressing to realize that even as our understanding of the importance of tropical ecosystems increases, tropical rainforests themselves are shrinking. If current trends continue, there will be nothing but relatively postage stamp-sized remnants.

To slow or even reverse such trends, conservation professionals use a variety of programs and policies ranging from narrowly tactical to broadly strategic. Scientists and managers draw from the collection of tools listed below to protect as much life as possible.

• Developing Captive Breeding Programs

Captive breeding programs at zoos are sometimes the last hope for endangered species that don't have a prayer in the wild. The San Diego Zoo, for instance, bred endangered California condors and saved them from extinction. Although important when it is the only option, captive breeding is relatively expensive and usually reserved for the few charismatic animals that get people excited. Unfortunately, it does not always succeed. Biologists, ornithologists, and veterinarians involved in the government's captive breeding program were unable to save the dusky sparrow, for example, and the last birds died in captivity.

BOX 4-1

STRATEGIES TO PROTECT BIODIVERSITY

- Developing Captive Breeding Programs
- Intervening in Genetics
- Protecting Species
- Removing Predators
- Protecting Habitats
- Restoring Habitats
- Passing Laws and Regulations
- Providing Economic Incentives and Removing Perverse Subsidies

- Intervening in Genetics

When the number of individuals in a species grows very small, the species tends to become inbred, threatening its long-term survival. Genetic intervention can help. This tool helped the Florida panther population survive when scientists with the U.S. Fish and Wildlife Service brought in panthers from Texas to improve the bloodline, critically enhancing their chances of survival because of greater gene variability.

- Protecting Species

Species protection is closer to the work that most conservation science students think about. Some species have needs that require unique actions, such as the remaining 315 right whales in Atlantic waters off the North American coast. These whales will likely survive if they produce more than nine young annually; however, generating fewer offspring threatens their existence. Each year the whales migrate from Florida to Massachusetts, directly through one of the busiest shipping lanes in the world. To protect the species, a team of scientists flies planes over them and also follows their progress in boats. **Geographic Information Systems (GIS) specialists** and other monitoring professionals from agencies like the National Oceanic Atmospheric Administration (NOAA) and contracted academic institutions track them twenty-four hours a day and send messages to ships to steer clear of the whales' migration.

- Removing Predators

Predator removal can help endangered plants and animals. Sometimes it involves a little weeding. We can see this in Hawaii, where introduced goats are being removed to protect plant species, or in the American West, where burros are the problem. Often, weeding must apply to actual weeds. Introduced plants take root and proliferate, threatening native species. For example, water hyacinth, a floating plant introduced from South America, can double in size in under two weeks. One of the worst weeds in the world, it reduces fisheries, shades out submersed plants, and seriously reduces biodiversity. Large areas of Lake Victoria in central Africa have become choked with hyacinths. Wildlife managers and botanists are introducing herbicides, insects to eat the plants, and machines that grind up its dense thickets as control measures to fight this problem.

GEOGRAPHIC INFORMATION SYSTEMS SPECIALIST

AT A GLANCE

Employment:

> 60,000 (includes cartography, remote sensing, and related work)

Demand:

> Excellent

Breakdown:

> Public sector, 70 percent
>
> Private sector, 30 percent

Trends:

> • The National Imagery and Mapping Agency alone expects to hire 7,000 GIS people from 2004 to 2007.
>
> • GIS skills are increasingly required by other professionals, slightly slowing demand for specialists.

Salary:

> Starting salaries begin at $29,000, with annual median earnings equal to $47,000. Most earn between $35,000 and $60,000, with top salaries over $75,000.

JOB DESCRIPTION

In the environmental sector, information is essential for decision making. Geographic Information Systems (GIS) specialists enable environmental professionals like foresters and land use planners to collect, analyze, and integrate data about a specific location from disparate sources. "It's the study, display, and analysis of geospatial data, or data that's tied to a location on the earth," says Emmor Nile, GIS coordinator for the Oregon Department of Forestry.

These specialists combine data with digital maps to produce multiple layers of geographically referenced information. For example, a GIS study that documents pollution control can combine information about the sources of pollution, contaminated areas, and environmental impacts into a single report.

Ken Cochrane, a GIS specialist working on a coastal resources management project in the Northern Mariana Islands, says his work on pollution control requires performing different tasks. For example, Cochrane creates maps that show the anticipated impacts of development on the quality of coastal waters in coastal development plans that must go through permitting and land use planning processes. Cochrane also updates maps and databases with new information, and manages staff members who use GIS and collect data to support pollution control.

GIS specialists analyze data and design, implement, and maintain GIS applications for various projects. They use specific GIS software and technology such as ArcView, ArcInfo, ArcGIS, ArcCAD, and Visual Basic. As this technology becomes more user-friendly, GIS specialists are starting to introduce such programs to the public. In this regard, GIS specialists can serve as public outreach coordinators. Overall, GIS specialists work at nonprofit organizations, government agencies, and private firms.

GETTING STARTED

• A bachelor's degree in GIS, geography, cartography, or computer science is sufficient for employment, while master's degrees offer advanced research opportunities.
• Most positions call for three to five years of previous experience with GIS technology. Some employers also require certification through national membership groups like the Urban and Regional Information Systems Association.
• Look for internships and job opportunities that introduce you to software specific to GIS. Experience with geographic positioning systems (GPS) is also helpful.

ADVICE FROM THE PROS

GIS Coordinator Emmor Nile entered the field as a former forest manager who learned GIS along the way. "My bias is toward hiring people with practical experience—someone with a geography or environmental science degree who learns GIS afterwards," he says. "College is not a place for education but a place to learn how to get an education." Nile also suggests gaining field experience in specific areas of interest, such as forestry, while pursuing a college education.

• Protecting Habitats

Habitat protection is the heart of biodiversity protection, and it's certainly conservation work that the public most clearly sees and avidly supports. Tactics for
protecting habitats include encouraging protection of privately owned lands
and setting up national and state parks and fish and wildlife refuges. It involves
buying and managing conservation lands or putting restrictions on land uses.
In aggregate, the results of these efforts are often called "protected areas." Simply securing the original protection, however, is only the beginning. We must
also manage these areas.

How does one decide where to place a protected area? By picking places
that have concentrated endemic species. This would mean setting up reserves
in those particular "hot spots" mentioned earlier when describing the unequal
distribution and rarity of species. **Conservation biologists** are working to
identify areas that, if protected, will have disproportionately large impacts on
global biodiversity, allowing us to best use limited resources.

As interest has grown in protecting threatened species, the number and
size of protected areas has also grown, in wealthy nations and everywhere
around the world. This is an undeniably good thing.

Even here, however, we have concerns, as some areas are protected only in
name. Lines may exist on a map, but inside boundaries, there are poachers
killing game, farmers and **foresters** clearing land, homes and roads being
built, even rivers being dammed.

Drawing boundaries of protected areas is critical. However, biodiversity
protection was not always prime selection criterion for national parks and
such. In Hawaii, for example, a volcanic park goes from sea level to mountaintop
in an hourglass shape. Geologists, rather than **conservation biologists**, orig-
inally designed this park, resulting in the most interesting species existing outside of the hourglass "waist." Fortunately, this has changed and now much scientific sophistication is involved in selecting and managing protected areas.

Although habitat protection happens mostly inside designated areas, we
must find ways to save habitats outside protected boundaries to be serious about
biodiversity. This means improving management of privately owned lands.

• Restoring Habitats

Once habitats are destroyed, the only remaining conservation tool is restoration. In North Branch, a suburb of Chicago featuring a once-protected strip of

CONSERVATION BIOLOGIST

AT A GLANCE

Employment:

Over 60,000 biologists work in environmental or conservation specialties.

Demand:

Good

Breakdown:

Public sector, 55 percent

Private sector, 35 percent

Nonprofit sector, 10 percent

Trends:

• Demand for qualified and well-educated "conservation biologists" is increasing as resource agencies embrace the profession's management focused skill base.

• Only a small fraction—perhaps 5,000—of the nation's fish, wildlife, and related biologists consider themselves "conservation biologists."

Salary:

Graduates with bachelor degrees earn entry-level salaries around $30,000 and M.S. holders start at $36,000. The median salary for environmentally focused biologists is $53,000 with top wages averaging in the high $70s.

JOB DESCRIPTION

Conservation biology emerged in response to the growing loss of biological diversity in the world. Conservation biologists' main objective is to determine how to reduce or eliminate threats to biodiversity and, when possible, restore the health of ecosystems and their species. "Conservation biologists focus on those aspects of natural resource management associated with the network of subjects affecting biodiversity, such as genetics, population biology, sociology, and economics," says Allan Thornhill, executive director of the Society of Conservation Biology (www.conbio.net) in Arlington, Virginia. "They use the best possible science to guide their conservation work."

Conservation biologists find their niche in many working environments. Some focus on fieldwork, collecting and analyzing data and monitoring habitats and wildlife compared with population growth. Others work for nonprofit organizations performing scientific research. They often simultaneously juggle administrative tasks such as grant writing, public speaking, budgeting, and fundraising. Those working as college professors usually research and write papers and give presentations about their work.

Conservation biologists specialize in specific fields like genetics, wildlife biology, and economics, and direct their work accordingly. Regardless of their specific expertise, they are all working to prevent plant and wildlife extinction. This requires a keen understanding of the biological sciences and an awareness of the different threats that challenge species and their habitats.

GETTING STARTED

• A bachelor's degree in conservation biology or biological sciences with concentrations in either environmental studies or social sciences is ideal.
• For research and academia positions, a master's degree is highly recommended. Some leading graduate programs include the University of Michigan's School of Natural Resources and Environment program and Cornell University's Environmental Studies program.
• Look for fieldwork experience, which offers opportunities to analyze data and learn computer skills.
• International experience or experience in remote locations is recommended, as conservation biologists often work all over the world.

ADVICE FROM THE PROS

Timothy H. Tear, director of conservation science for The Nature Conservancy's Eastern New York chapter, advises that aspiring conservation biologists hone their writing skills, learn public speaking, and study environmental and biological sciences. Above all else, prospective professionals should learn statistics and Geographic Information Systems (GIS). "If there's one skill that will help you understand conservation biology literature, it's gaining a solid understanding of statistics," he says. "The use of spatial information and analysis is becoming increasingly more important in the field. These are skills each student emerging today should have."

land that is now abandoned to weeds, Steve Packard, a restoration ecologist, has led an effort to bring back part of the original oak savannah ecosystem. This is in the heart of a huge metropolitan area, not to mention that natural ecosystems cover less than 0.1 percent of Illinois.

The first restorative step was to clear the weeds and then replant seeds for native plants to grow. However, initial efforts to plant native species failed. Animals ate the small, transplanted plants. Without fire, it would remain so. Fire posed several problems, but the group was undeterred; they burned their reserves and the results were immediate and dramatic. Prairie plants flourished. Weeds retreated except under native oak trees, which volunteers then replaced with local flora described in an 1846 plant list for the area. This approach illustrates two rules of ecological restoration. First, one needs to save all species needed to restore an ecological community. Second, one needs to know which species belong where.

Restoring larger portions of our lost prairies may not be so far-fetched. Long ago, much of North America was unending miles of prairies. This was virtually all destroyed and transformed into farms and towns. Now, however, people have been leaving the land in large numbers, driving down land prices so that we can at least imagine creating a large prairie land national park, complete with buffalo.

- Passing Laws and Regulations

Enacting regulations that create institutional frameworks to protect nature is a crucial tool for protecting biodiversity. Legislative and **policy analysts** play an important role in crafting laws and regulations on both domestic and international fronts. Among the most important laws already in place in the United States are the Endangered Species Act, the Marine Mammal Protection Act, and the National Forest Protection Management Act. Internationally, we can point to United Nations-sponsored treaties and collaborations as having great impacts, such as the Convention on Biological Diversity and the Convention on International Trade in Endangered Species of Wild Fauna and Flora (CITES). For example, CITES affords protection in varying degrees to more than 30,000 species.

- Providing Economic Incentives and Removing Perverse Subsidies

Assigning economic value to biodiversity is necessary but challenging. Our natural environment is often taken for granted. Providing incentives and

Career Spotlight

The Center for Applied Biodiversity Science at Conservation International Washington, D.C.

CABS scientist Leeanne Alonso searches for insects as part of a rapid response biological assessment in Brazil.

With 1,000 staff people in thirty-four countries, Conservation International (CI) is one of the largest and most effective environmental organizations in the world. In collaboration with governments, academic institutions, and others, CI employs many biodiversity protection strategies in its work.

The scientific and technical resources of CI are consolidated within the Center for Applied Biodiversity Science (CABS), making it a perfect spotlight selection for exploring some critical biodiversity careers.

"We have eighty professionals with the Center, and about 50 percent are biologists and ecologists, many at the Ph.D. level, reports CABS director Kristen Walker-Painemilla. "Twenty-five percent of the staff are people with training in GIS, remote sensing, database management, and website work. Approximately 15 percent are economists, anthropologists, and other social scientists. The rest are administrators, communicators, fundraisers, and finance people."

Life for CABS biologists is exactly as fascinating as one might imagine, according to Leeanne Alonso, senior director of the Center's Rapid Assessment Program (RAP). Under Alon-so's leadership, it's the job of expert scientists working in the RAP unit to rapidly collect biological information in unexplored tropical areas before it's too late. RAP teams have discovered hundreds of previously unknown plant and animal species from Papua New Guinea to Peru.

Alonso's career path is a good template for aspiring scientists with an interest in biodiversity protection. "My B.A. degree was in liberal arts, and I followed that with an M.S. in zoology, and a Ph.D. in biology from Harvard," she says. "My research focused on tropical ecology and ants, which was great scientific preparation for the CABS focus on the tropics."

Although her research skills and knowledge of tropical ecology are both essential on the job, Alonso notes that coordination of people, money, and logistics are just as likely to dominate her day. "The ability to form teams and manage people is especially crucial for this work."

Scientists like Alonso are central players in biodiversity protection work, but information technology professionals are not far behind, according to Daniel Juhn, director of the GIS Mapping Laboratory at CABS. "Mapping and monitoring biodiversity is a prerequisite to protecting it," Juhn explains. "We have a strong need for GIS and remote sensing professionals to do that."

Sonya Krogh is a GIS specialist and Karyn Tabor a remote sensing specialist on Juhn's team. They both agree that the future is bright for people with their skills. With a M.S. degree in applied geography from New Mexico State University in 2000, Krogh is currently working on transforming data from the mountain regions of Peru into an interactive GIS that others will use to make conservation decisions.

"Clearly, technical competence with advanced GIS computer programs is essential for this job," Krogh explains. "You have to love exploring the technology." At the same time, she believes that GIS work is ultimately about serving people, even if all they need is a simple map.

Karyn Tabor rarely travels to the hot spots that CI protects, but she knows them intimately. "Remote sensing can feel like science fiction," Tabor says. "Through a combination of satellite tracking, high-tech sensors, and nanotechnology, we can imagine that we might someday monitor environmental changes remotely in real time."

Tabor earned a master's degree in remote sensing/GIS from Boston University in 2001 through a five-year program that also involved a B.S. in environmental studies. She says she needs all of this training in order to be competitive for her job. "The quality of remote sensing degree programs has improved dramatically. This has become a demanding specialty."

Computerized GIS informed by data from scientists on the ground and from remote sensing technology is helping CI achieve improved results. Making this information accessible over the Internet will widen the circle of learning beyond CABS to others around the world.

That's where Karen Desmond comes in. As a web applications developer in CABS "Conservation Knowledge" department, Desmond is finishing an application that will allow students around the world to track species gains and losses online for regions across the planet. With luck, thousands of future biodiversity protection professionals and advocates will be using it tomorrow.

removing perverse subsidies that discourage or prohibit conservation efforts must be part of every conservation biology strategy. For example, Costa Rica's government pays to protect forest watersheds in order to slow the rate of deforestation. The city of New York has improved its water supply by protecting watersheds, instead of greatly expanding water treatment plants. These decisions proved to be both cost-effective and conservation-minded.

Removing harmful subsidies is equally important. Environmental scientist Norman Myers and researcher Jennifer Kent suggest a powerful conservation strategy in examining the "sensibility" of subsidies, given concerns about the environment. Assessing current subsidies to the fishing industry highlights the need for subsidy removal. The total amount of the world fisheries catch is estimated at around fifty billion dollars. That's a lot of money for the fishing industry, but we spend more than twice that to catch those fish. Massive subsidies cause incredible damage to oceanic life and coastal biodiversity, and, therefore, we actually provide incentives to harm nature through our fisheries policies. Similar patterns in agriculture, forestry, and elsewhere also exist. These patterns demonstrate immense challenges that people working to preserve biodiversity must overcome in order for their conservation efforts to be effective.

Career Advice from the Expert

Conservation is a crisis discipline, one demanded by the unusual rates of loss. It's also a mission-driven discipline. By analogy, ecology and conservation have the same relationship as physiology and medicine. Physiology studies the workings of the human body; medicine is mission-oriented and aims to understand what goes wrong and how to fix it.

Similarly, conservation science is about urgently solving problems, making a **conservation biologist's** work differ from scientists working in wildlife or fisheries biology, or evolutionary ecology. Conservation science is also about exploring. The field's evolution has reflected our change in values. When I first started, forestry, fisheries, and wildlife professionals focused mainly on timber and game. There was much talk about "multiple use," but multiple use still meant that you chopped down trees. Now, however, we find that people—not only scientists—really want forests to be forests.

These types of changes have been reflected in our colleges and universities. The Yale School of Forestry and Environmental Studies, the Nicholas School of the Environment at Duke University, the Center for Environmental Research and Conservation at Columbia, and Stanford University are among the best schools for the biodiversity protection field.

Students and recent graduates must gain field experience to be competitive. To be serious, it is important to travel to the places where biodiversity loss is greatest and the work most urgent. Going to these places for pleasure also increases the appreciation of such resources. It will help to have technological skills like GIS and remote sensing as well.

Overall, there is modest employment growth in this field compared to the work that must be done—we still don't know the names of nine out of ten species on this planet! There are an extraordinary number of ways to help, and protecting biodiversity is remarkably rewarding because everyone involved is very passionate.

RESOURCES

Association for Biodiversity Information, www.aib.org

Center for Applied Biodiversity Science, http://www.biodiversityscience.org

Conservation International, www.conservation.org

Everglades Restoration Plan Organization, www.evergladesplan.org

Global Biodiversity Forum, www.gbf.ch/

IUCN/The World Conservation Union, www.iucn.org

National Biological Information Infrastructure, www.nbii.gov/issues/biodiversity/

NatureServe, www.natureserve.org

Society for Conservation Biology, www.conbio.net/scb

The Nature Conservancy, www.nature.org

U.S. Fish and Wildlife Service, www.fws.gov

5. CLIMATE CHANGE

Eileen Claussen (standing), president of the Pew Center on Global Climate Change, at a Pew workshop with with Nobel Prize winner Dr. Joseph Stiglitz (right) of Columbia University and Nobel Laureate Dr. Kenneth Arrow (left) of Stanford University.

A CONVERSATION WITH
Eileen Claussen

Eileen Claussen, president of the Pew Center on Global Climate Change, is a leading voice for sensible action to address climate change—the most pressing global environmental problem of the twenty-first century.

"I am realistic about the challenges we face on this issue. There are still a small number of skeptical scientists and others who prefer to do nothing about climate change," Claussen says. "And some of these people have loud megaphones. But I do not believe they will prevail. I believe strongly that there are many certainties in science; that there are many actions we can take with no negative economic impact; and that, with careful planning and continued technological development, we can address this problem and still grow the global economy. Which is precisely what we need to do."

Before becoming president of the Pew Center, Claussen served as the Assistant Secretary of State for Oceans and International Environmental and Scientific Affairs. She also has served as Special Assistant to the President and Senior Director for Global Environmental Affairs at the National Security Council, and spent over twenty years at the U.S. Environmental Protection Agency. Claussen received her M.A. degree in English from the University of Virginia.

What Is the Issue?

The greenhouse effect is a naturally occurring process that warms our planet. In short, certain gases, like carbon dioxide (CO_2), methane (CH_4), and nitrous oxide (N_2O), act like the roof of a greenhouse—letting the sun's light in and trapping heat. Without the greenhouse effect, the earth would be approximately 60 degrees Fahrenheit colder. While most atmospheric greenhouse gases (GHGs) have both natural and anthropogenic (or human-induced) sources, anthropogenic activities—everything from industrial processes to fossil-fuel combustion to changes in land use, including deforestation—have significantly increased GHG emissions into the atmosphere. These emissions are upsetting the balance between natural GHG sources and natural sinks—processes that result in the net removal of GHGs from the atmosphere. As a result, more GHGs are being emitted than are being removed, and concentrations in the atmosphere are rising, trapping more heat and increasing temperatures.

Scientific records from global temperature readings over the last century leave no doubt that the earth is indeed warming. Surface temperatures have risen about 1 degree Fahrenheit since the late 1800s. The most recent estimates by the Intergovernmental Panel on Climate Change (IPCC), a United Nations body that draws on the expertise of hundreds of climate scientists around the world, indicate that, under a "business as usual" scenario, average global temperature will rise 2.5 to 10.4 degrees Fahrenheit (1.4 to 5.8 degrees Celsius) by the end of the twenty-first century. (See IPCC, *Climate Change 2001: The Scientific Basis,* for more details.) This is a significant change: the high end of this range is equal to the change in the average global temperature associated with the end of the planet's last ice age 10,000 years ago. But during that ice age it took thousands of years to reach this level of warming—not just one century.

The U.S. National Research Council of the National Academy of Sciences recently confirmed in a report to President George W. Bush the IPCC's finding that "greenhouse gases are accumulating in earth's atmosphere as a result of human activities, causing surface air temperature and subsurface ocean temperature to rise." (See the National Research Council's full report, "Climate Change Science: An Analysis of Some Key Questions," for more findings.) Potential consequences of this warming include sea-level rise and increases in the severity or frequency (or both) of extreme weather events, including heat waves, floods, and droughts, with potentially major impacts to U.S. water resources, coastal development, infrastructure, agriculture, and ecological systems. The risks of these and other consequences are sufficient to justify action to significantly reduce GHG emissions.

Reducing GHG emissions is particularly difficult because no single country or industry is solely responsible for the problem. GHGs are emitted from a range of anthropogenic activities across all economic sectors and mix uniformly in the atmosphere, where they have long lifetimes. Simply put, a ton of GHGs emitted in the United States has the same impact on the climate as a ton emitted in Malaysia. This makes climate change a quintessential collective action problem—a global challenge that, in the long run, will require worldwide collaboration in order to avoid serious consequences. Nonetheless, the United States is responsible for about 25 percent of GHG concentrations in the atmosphere to date, so we have a special responsibility to lead the world in addressing the issue.

How Are Environmental Professionals Approaching the Issue?

A response to the global climate change challenge must begin now if it will be effective, and it must include both short- and long-term components. While an effective solution must be global, it is important that the United States be mindful of its unique role and responsibilities—both as the world's largest producer of GHGs and as a leader within the world community.

Establishing a clear path for GHG emissions reductions would begin this timely and efficient response. We can take certain steps now; for example, there are countless ways to use energy more efficiently and thereby reduce GHG emissions. Ultimately, we must fundamentally transform the way we power our homes, factories, and cars—in short, the way we power our entire economy. In this, we should remember the words of Eleanor Roosevelt: "The future is literally in our hands to mold as we like. But we cannot wait until tomorrow. Tomorrow is now."

• Creating an Alternative Energy Economy

Current GHG emissions consist primarily of CO_2 from the combustion of fossil fuels in electricity generation, buildings, industrial processes, and transportation activities. To achieve the GHG emissions reductions necessary to address climate change, we must steadily reduce our dependence on fossil fuels and

BOX 5-1

ADDRESSING CLIMATE CHANGE

- Creating an Alternative Energy Economy
- Designing and Implementing Public Policy
- Understanding the Kyoto Protocol
- Rethinking U.S. Climate Change Policy
- Acting at the State and Local Level
- Engaging Business and Industry

develop new energy sources that have fewer or no emissions in order to create an alternative energy economy that is more climate-friendly.

In the near term, increasing the use of natural gas in lieu of coal or oil can decrease GHG emissions, because natural gas is the least carbon-intensive of these fuels. Finding ways to capture and sequester, or remove, GHG emissions from the atmosphere—particularly from coal combustion—so that they are not released into the atmosphere can also help address climate change even as we find less carbon-intensive ways to produce energy. Researchers in the field of geochemical engineering are developing these technologies today, and a few pilot projects have begun. For example, American Electric Power has announced that its Mountaineer plant in West Virginia will be the site of a $4.2 million carbon sequestration research project funded by the U.S. Department of Energy and a consortium of public- and private-sector participants. Projects like this are a start, but in the longer term, further development of renewable energy sources such as wind, solar, geothermal, hydrogen, and biomass could be critical to achieving the GHG reductions necessary to address climate change. For more information about alternative energy sources, see chapter 9 about energy.

The U.S. transportation sector is also heavily reliant on fossil fuels, currently accounting for one-third of U.S. CO_2 emissions. U.S. transportation produces more CO_2 emissions each year than any other nation's entire economy, except China. (See Green and Schaefer, *Reducing Greenhouse Gas Emissions from U.S. Transportation*, for more information.) Current vehicle trends contribute to the emissions problem: many people are driving less efficient vehicles due to the increasing popularity of sport utility vehicles, and demand is growing for automobiles in developing countries where there are fewer vehicles. In this regard, **transportation planners** and analysts are challenged to find ways to provide people with mobility in climate-friendly ways.

In the near term, hybrid vehicles (where an internal combustion engine is complemented by an electric motor) can provide important gains in fuel economy. As reported in *Reducing Greenhouse Gas Emissions from U.S. Transportation*, by making the most effective use of both power sources, the advanced hybrid design in combination with a continuously variable transmission can improve fuel economy by 40 to 50 percent. Hybrid vehicles are already commercially available in the United States from Toyota and Honda, and other manufacturers like Ford, GM, and DaimlerChrysler have announced plans to introduce hybrids in the next few years. Due to the increasing popularity of this technology, the design and construction of new generations of hybrid vehicles is an up-and-coming professional niche among engineers in the automotive industry.

In the long term, fuel cell vehicles are a promising alternative to vehicles that run on petroleum. Emissions from hydrogen-powered fuel cell vehicles consist mainly of water vapor. However, in order to shift to fuel cell vehicles, we must overcome technological challenges like determining hydrogen sources. Hydrogen production from natural gas or coal (but only with CO_2 capture and sequestration) is a reasonable beginning, but ultimately renewables must become the source of hydrogen in order to significantly reduce the transportation system's carbon intensity. We would also have to establish the appropriate infrastructure to support hydrogen-fueled transportation and replace traditional gas stations, tanks, and pipelines. Despite these challenges, some manufacturers are now producing test fuel cell vehicles in order to advance this promising technology. To effect these important changes, research and development professionals' efforts must be complemented by business leaders who can help create supply and demand for climate-friendly products, practices, and markets. For more information about reducing emissions and efficient transportation, see chapter 9 on energy.

- Designing and Implementing Public Policy

Well-crafted international and domestic public policies are central to reducing GHG emissions effectively. Such policies cannot be well designed without input

CLIMATE CHANGE SPECIALIST

AT A GLANCE

Employment:

2,000

Demand:

Good

Breakdown:

Public sector, 60 percent

Private sector and academia, 40 percent

Trends:

• Climate change specialists represent a growing specialty within the nation's 8,000 atmospheric scientists, meteorologists, and related fields.

• Demand is increasing as research continues to point to dangerous changes resulting from global warming.

• The National Oceanic and Atmospheric Administration (NOAA) is a prominent employer in the field.

Salary:

Median salaries are just over $61,000. Starting wages are $31,000 for those with bachelor's degrees and $44,000 for master's graduates. Most specialists earn between $41,000–$79,000, but top earners average over $94,000 annually.

JOB DESCRIPTION

As a result of increased greenhouse gas emissions, scientists worldwide warn of profound changes resulting from dramatic climate change. Climate change specialists measure the climate and how it changes over time to decipher what causes global warming, widespread species loss, coastline erosion, and other dangerous effects to the environment. Dr. Warren Washington, senior scientist at the National Center for Atmospheric Research (www.ncar.ucar.edu) in Boulder, Colorado, says, "We sort out the myriad relationships between climate and those factors which are affecting the global temperature."

Many climate change specialists meet with colleagues to process and analyze data generated by computer-powered models that mimic interactions between climate and increasing greenhouse gas emissions. For example, specialists can calculate changes in sea levels using satellite and radar technology to determine if water is rising. Since water expands when it heats up, higher sea levels reflect a warmer atmosphere and point to global warming.

Climate change specialists often present their research at lectures and in written reports. This type of research can lead to meaningful changes in international and national policies concerning greenhouse gas emissions and their impacts on the natural environment. The best researchers have opportunities to influence and participate in the Intergovernmental Panel on Climate Change (www.ipcc.ch) every five years, an international body that assesses climate change research and makes recommendations to elected officials and policy makers who influence global climate change standards.

Many climate change specialists work as atmospheric scientists, meteorologists, or climatologists. They can work for government agencies and academic institutions, and alternatively in the private sector as weather forecasters or at private consulting firms, where they often compile and analyze climate data for various industries. "We apply a multidisciplinary approach," says Randy Fried, senior vice president for ICF Consulting in Fairfax, Virginia. "We also involve wide-ranging professionals, from foresters to chemical engineers, to analyze the climate change problem, which is clearly happening."

GETTING STARTED

• A bachelor's degree in atmospheric science and meteorology is highly recommended. Physics, mathematics, chemistry, and engineering are excellent fields for graduate studies.

• For advanced research positions, a Ph.D. is preferable.

• Seek fellowships, research assistant positions, or internships that will introduce you to working with computer-aided models and global positioning systems.

ADVICE FROM THE PROS

Michael Wyllie, a leading meteorologist at the National Weather Service in Upton, New York, says his work involves tracking future climate change sources. "We're the ones collecting all the data. It's becoming much more important as we get into the twenty-first century and start noticing the effects on the environment." He advises that students learn all types of computer systems, especially those used in climatology and weather research, through internships at weather facilities like the National Weather Service or state weather agencies.

from **climate change specialists**, who research both the causes and the consequences of those emissions. Because climate change is a global challenge that requires a global solution, policy experts—including **elected officials**, **lawyers**, economists, environmental scientists, **public policy analysts**, **lobbyists**, and international negotiators—must start with a full understanding of how the issue is being addressed internationally before they can design successful programs at home.

International efforts to address climate change began with the 1992 United Nations Framework Convention on Climate Change (UNFCCC), which set an ultimate objective of "stabilization of greenhouse gas concentrations in the atmosphere at a level that would prevent dangerous anthropogenic interference with the climate system." However, the voluntary emission reduction targets set by the UNFCCC were not sufficient and would not be met, leading to negotiation of the 1997 Kyoto Protocol.

· Understanding the Kyoto Protocol

The Kyoto Protocol is an international treaty that would establish binding emissions targets for industrialized countries and a range of mechanisms to encourage cost-effective compliance. The emissions targets for the period 2008–2012 average 5.2 percent below 1990 levels and range from 8 percent below (in the European Union) to 10 percent above (in Iceland). The U.S. target negotiated by the Clinton Administration is 7 percent below 1990 levels. The protocol will enter into force when ratified by at least fifty-five parties to the convention, including industrialized countries whose emissions account for at least 55 percent of the total CO_2 emissions in 1990 from that group.

President George W. Bush has made clear he will not submit the Kyoto Protocol to the U.S. Senate for ratification. Nevertheless, other countries continue to support the protocol and as of November 2003, 119 countries, accounting for 44.2 percent of global emissions, had ratified or acceded to the agreement. The European Union, for example, has implemented a CO_2 emissions trading scheme designed to help meet its Kyoto commitments. The remaining hurdle for the protocol is the 55 percent threshold mentioned earlier. If Russia should ratify the treaty, for example, this hurdle would be met and the Kyoto Protocol would go into effect even without U.S. ratification.

Implementing the Kyoto Protocol would be a strong signal to markets that GHG emissions come with costs and a declaration of multilateral will to confront this global challenge. But against that challenge, Kyoto would be only a first step. Whether or not Kyoto comes into force, we must begin to look beyond it. Several core challenges remain, such as establishing practical goals beyond the initial targets set by Kyoto, determining a fair way to include developing countries in climate change mitigation efforts, and ensuring that efforts to address climate change are as cost-effective as possible.

- ## Rethinking U.S. Climate Change Policy

As the world's largest economy and the world's largest GHG emitter, the United States is central to any long-term strategy to address global climate change. Voluntary efforts in a number of sectors over the past several years have failed to curb overall growth in U.S. GHG emissions, which rose roughly 12 percent over the past decade and are projected to continue rising for the foreseeable future. To be effective and affordable, a long-term emissions reduction program must couple mandatory GHG reductions with technology development and market mechanisms.

Ideally, a national strategy would be guided by a specific long-term emissions goal. It would also couple short- and long-term measures—and both supply and demand elements—to signal markets to begin the transition toward that ultimate objective. More specifically, short-term measures are needed to improve energy efficiency and encourage the use of lower-carbon fuels. Long-term measures are needed to encourage sustained investment in developing the technology and infrastructure needed to facilitate the transition to a low-carbon economy.

A domestic strategy ultimately must reflect our country's international commitments. However, its design and implementation should proceed now, even if the United States is not yet prepared to enter into an international agreement. As domestic and international programs evolve, close coordination between them is critical. This is especially important for companies that operate, compete, and sell products both domestically and abroad, as they will be subject to rules dealing with climate change in other countries. In addition, coordination is necessary to maximize the effectiveness of emissions trading and other flexibility mechanisms that are being developed at the international level.

Although the United States has not yet implemented a mandatory national climate change policy, interest in climate change proposals in the U.S. Congress has increased. Members of both political parties are more willing to demonstrate their interest in climate protection. The number of climate change-related bills introduced has risen dramatically from seven in the 105th Congress (1997–1998) to over 80 in the 107th Congress (2001–2002). Perhaps most significantly, on October 30, 2003, the Senate voted on the Climate Stewardship Act (S.139), a bill crafted by Senators John McCain (R-AZ) and Joe Lieberman (D-CT). The bill couples strong environmental goals with a flexible market-based approach that allows business to reduce GHG emissions at the lowest possible cost. While the bill did not pass, the close vote (55–43) and debate that preceded it showed growing bipartisan support for real action against climate change. Advocates and **lobbyists** are working to produce an even better result the next time around.

- ## Acting at the State and Local Level

Despite the absence of a comprehensive climate policy at the national or international level, U.S. states and local communities are surprisingly active on this issue. States are able to address climate change through their authority over land use, transportation, utilities, taxation, and other policy areas affecting the environment. At the local level, cities are adopting climate change policies and engaging citizens with programs that encourage changing lighting practices or planting trees. Indeed, state and local efforts illustrate that climate change can be a bipartisan issue, an economic development opportunity, and an opportunity for policy entrepreneurship.

Twenty-seven states have developed or are developing strategies or action plans to reduce net GHG emissions. For example, New Jersey has committed to reduce GHG emissions to 3.5 percent below 1990 levels by 2005. New England governors signed an agreement in August 2001 with the Eastern Canadian premiers for a comprehensive Climate Action Plan that aims to reduce overall emissions and set targets on a regional level. New Hampshire, Massachusetts, and Oregon have set emissions requirements for power plants. Texas and Minnesota require that a specific amount of new electricity generating capacity be based on renewable energy. Furthermore, states are reducing agricultural and transportation sector emissions and promoting energy efficiency. However, it is important to note that state and local actions are not substitutes for comprehensive national or international approaches.

• Engaging Business and Industry

Business and industry must be engaged in shaping and implementing climate solutions because their operations are major sources of GHG emissions. Companies can demonstrate leadership by establishing and meeting emissions reduction objectives; investing in new, more efficient products, practices, and technologies; and supporting action to achieve cost-effective emissions reductions.

The Pew Center has identified more than forty companies, most either based in the United States or with significant U.S. operations, that have voluntarily committed to GHG reduction targets. British Petroleum, for example, has reduced GHG emissions to 10 percent below 1990 levels—eight years ahead of target—and now has pledged to keep them there at least until 2010. Alcoa is working to reduce its GHG emissions by 25 percent below 1990 levels by 2010. DuPont has achieved a 65 percent reduction below 1990 levels, well ahead of its 2010 target, and has pledged to hold emissions flat. (See Margolick and Russell, *Corporate Greenhouse Gas Reduction Targets*, for more details.)

The Pew Center recently studied several companies that have taken on targets and found they have multiple motivations. They believe the science of climate change is compelling. They know in time the public will demand strong climate protections, and they can get ahead of the curve by reducing their emissions now. They want to encourage government policies that will work well for business. They also want to improve their competitive position in the marketplace. And that, in fact, has been the result. Companies are finding that reducing emissions also helps improve operational efficiencies, reduces energy and production costs, and increases market share—all things that contribute to a healthier bottom line. While addressing climate change is not necessarily profitable, the evidence so far suggests it is certainly affordable.

Most companies have environment, health, and safety staff that work to implement such emissions reductions. Accountants and other financial professionals determine the costs and benefits of implementing such programs; mechanical and electrical engineers and technical professionals redesign projects, production processes, and operating procedures. Additional jobs are being created for brokers and auditors who will implement and evaluate GHG emissions trading programs. Many of these professionals work with public relations staff that communicate corporate-wide reduction strategies to the government, public, and company employees. As a growing number of companies

LOBBYIST

AT A GLANCE

Employment:

5,000

Demand:

While no firm numbers exist to document the demand for lobbyists, the American League of Lobbyists reports that demand has remained steady over many years.

Breakdown:

Private, 50 percent

Nonprofit, 50 percent

Trends:

• Increased assaults on the environment have heightened the need for lobbyists working on behalf of our natural resources.

• Lobbyists are often the first to lose their positions during economic recessions.

• Competition remains keen for entry-level positions in this popular field.

Salary:

Starting salaries range from $20,000 to $35,000. Experienced lobbyists earn between $50,000 and $80,000, with the top salaries regularly exceeding $100,000.

JOB DESCRIPTION

Environmental lobbyists pressure state and federal lawmakers to vote for pro-environment initiatives that are under review. Many lobbyists work in Washington, D.C., for nonprofit organizations and public advocacy groups looking to bridge the gap between Capitol Hill and the public.

Lobbyists represent their organizations by meeting and debating with politicians, opponents, allies, and other stakeholders. Successful lobbyists base much of their work upon building and maintaining strong relationships with multiple stakeholders. When their employers cannot make large campaign donations to politicians, lobbyists often collect money from other sources or campaigns.

Grassroots lobbyists enlist the help of local residents to influence politicians or demonstrate on behalf of their organization. This equates to long hours writing letters, holding press conferences, and rallying for community support. "As a lobbyist, I'm harnessing the power of the grassroots to exert pressure on the political process," says Karen Wayland, a lobbyist for the Natural Resource Defense Council.

Lobbyists must be well-informed and have an in-depth understanding of pressing environmental issues. Effective lobbyists are persuasive and well-spoken debaters who build reputations for being unwilling to back down from their political stances. They must also remain aware of their opponents' positions, as well as the pressures that lawmakers are under and the reality of current political tides.

On a daily basis, lobbyists attend committee hearings on Capitol Hill, research campaigns, write press releases, and meet with lawmakers and the media to advocate their political views. This intense career can be extremely rewarding when a pro-environmental bill is passed; however, lobbyists also face stress and frustration when a bill is defeated.

GETTING STARTED

• For most positions, a master's or law degree is highly recommended.
• While no specific area of study is required for lobbyists, political science, journalism, law, economics, government, and environmental studies are helpful for prospective lobbyists.
• Most positions require three years of postcollege work.
• Working for advocacy groups or nonprofit organizations can serve as a stepping-stone for later work in the field.
• Seek jobs or internships that will give you the chance to work on Capitol Hill.

ADVICE FROM THE PROS

Deanna Gelak, president of the American League of Lobbyists (www.alldc.org), says newcomers to this field must expect to work long hours and experience high levels of stress. "Get experience working on Capitol Hill early in your career," she says. "Executive branch or state and local government experience is also helpful. This provides insight into the process that you can't learn in a textbook. Concise writing skills are also critical."

Career Spotlight

National Center for Atmospheric
Research
Boulder, Colorado

In our time, global climate change has evolved from an unsettling scientific theory to a universally recognized international crisis. Along the way, the National Center for Atmospheric Research (NCAR) has been both an observer and an instigator of our emerging scientific awareness. "It's almost as if this whole arena of science is being born as we watch," notes director Tim Killeen, "and NCAR is right at the leading edge of it."

Founded in 1960 under the primary sponsorship of the National Science Foundation, NCAR is recognized for research initiatives in atmospheric chemistry, meteorology, solar and solar-terrestrial physics, and climate. Beyond these disciplinary studies, the center's researchers develop climate system models, which simulate the complex interactions among climate, weather, the sun, and the biosphere and oceans. Thanks to the power of supercomputers and the increased scope of these models, we know more than ever about the impact of human-emitted greenhouse gases on the world's climate. "We're much more sure of the science and the anthropogenic effects," Killeen says. "Scientists have made major strides."

Spread over two major sites in Boulder, Colorado, NCAR employs over 400 people in nine scientific divisions, including nearly half at the doctoral level, on an annual budget of $62 million. As one of the world's most respected research institutes investigating what is arguably the planet's most pressing environmental concern, NCAR is well-positioned to attract and support top international talent.

For example, Andrew Gettelman came to NCAR's Astmospheric Chemistry Division (ACD) with a Ph.D. in atmospheric sciences from the University of Washington and experience working in policy for Climate Action Network and the Natural Resources Defense Council—two leading nongovernmental organizations. Today, Gettelman works within the Climate and Global Dynamics (CGD) division, researching water vapor and the radiation balance between the upper troposphere and lower stratosphere, with the goal of improving climate models for this region. "Water vapor affects the chemistry and climate of these atmospheric layers," Gettelman explains. "Its study is important for understanding the greenhouse effect in climate change."

He also works with the Environmental and Social Impacts Group (ESIG) on the information gap in climate sciences between developed and developing nations, an outreach project he developed when he came to NCAR in 1999 as an Advanced Studies Program (ASP) postdoctoral fellow.

Another ASP fellow, Mausumi Dikpati, joined NCAR after receiving her Ph.D. in solar physics from the Indian Institute of Science in

NCAR scientists use networks of automated surface weather stations like this one to study weather patterns and air quality nationwide.

Bangalore, India. As a member of the High Altitude Observatory (HAO) Solar Interior and Variability group, Dikpati's research focuses on solar modeling.

"I'm trying to improve our understanding of the cyclical nature of solar magnetic features," she says, "because it has a profound influence on earth systems." A major contribution in her collaborative research shows that the flow of electrically charged plasma across the sun's meridians governs the critically important eleven-year solar activity cycles.

Understanding global climate change means studying the ocean as well as the atmosphere. Joanie Kleypas is a marine scientist in ESIG who specializes in the response of coral reefs to climate change. Much of her research has focused on the implications of dissolved carbon dioxide in seawater, which reduces the saturation level of calcium carbonate, the building block of reefs. As the saturation level declines, corals and calcareous algae accumulate less calcium carbonate in their skeletons. As a result, reef building may slow or stop, and the entire ecosystem could become far more vulnerable to erosion or other threats.

Kleypas has a master's degree in marine science and ecology from the University of South Carolina and a Ph.D. from James Cook University in Australia. She has been at NCAR since 1993, first as an ASP postdoctoral fellow and then as an associate scientist in CGD's oceanography section.

She also studies the ecological impact of warming ocean waters, which have been blamed for weakening or killing tropical reefs. She plans to expand her research to look at the impact of climate change on other coastal ecosystems, such as estuaries.

With such talented people on board, Director Killeen is optimistic about new breakthroughs in climate science. "The caliber and diversity of our fearless young scientists brings new vitality and scientific leadership," Killeen says with pride.

take steps to address climate change, job opportunities in these areas are likely to increase.

Career Advice from the Expert

Climate change is a complex issue, and solutions and careers will originate in multiple sectors. Governments, businesses, advocacy organizations, international negotiators, scientists, economists, **journalists**, and many others will contribute to climate change policy and solutions. We need experts in all of these fields in order to design and implement emission reduction programs. While various actors approach the issue from different angles and with different expertise, they must all share a commitment to the issue and to finding solutions. Because this is a relatively new and multidisciplinary issue, there is no traditional educational path toward a career in addressing climate change.

The Pew Center employs **policy analysts**, scientists, economists, **lawyers** (both domestic and international), and communication professionals, and while each brings a skill set that is different, all contribute equally. My background in liberal arts provided writing, speaking, and analytical skills. These skills have served me well in various positions over the years. Because climate change is complex and cuts across virtually all sectors of society, cultivating the ability to think creatively, ask the right questions, and maintain an open mind is key. It's also helpful not to be intimidated by technical or scientific jargon or detail.

Solid education, job experience, and an ability to deal with complex issues are important for success working in this field. The capacity to persevere in the face of difficulty is, however, absolutely crucial. Addressing climate change essentially requires a new industrial revolution, one in which we transform the way we power our society. Even under the best of circumstances, when policy and politics are in alignment, this is not an easy task. What this task definitely requires is an ability to see small accomplishments for what they are: steps on the path to a sustainable future. Only optimists should apply for jobs in this field.

RESOURCES

Carbon Dioxide Information Analysis Center, http://cdiac.esd.ornl.gov
Emissions Database for Global Atmospheric Research, http://www.rivm.nl/en/milieu/
Intergovernmental Panel on Climate Change, http://www.ipcc.ch/
International Energy Agency, Greenhouse Gas Programme, www.ieagreen.org.uk

International Institute for Applied Systems Analysis, www.iiasa.ac.at

National Center for Atmospheric Research, http://www.ncar.ucar.edu/ncar/

Pew Center on Global Climate Change, http://www.pewclimate.org

U.S. Department of State,
   www.state.gov/www/global/global_issues/climate/index.html

U.S. Environmental Protection Agency, www.epa.gov/globalwarming/

U.S. Global Change Research Program,
   http://www.usgcrp.gov/usgcrp/new.htm#variabilityandchange

U.S. National Assessment, http://www.gcrio.org/NationalAssessment/

United Nations Framework Convention on Climate Change, http://unfccc.int/

REFERENCES

Green, David L., and Andreas Schafer, *Reducing Greenhouse Gas Emissions from U.S. Transportation* (Arlington, VA: Pew Center on Global Climate Change, 2003).

Intergovernmental Panel on Climate Change, Working Group I, *Climate Change 2001: The Scientific Basis* (Cambridge, UK: Cambridge University Press, 2001).

Margolick, Michael, and Doug Russell, *Corporate Greenhouse Gas Reduction Targets* (Arlington, VA: Pew Center on Global Climate Change, 2001).

National Research Council, Committee on the Science of Climate Change, "Climate Change Science: An Analysis of Some Key Questions," 2001. Washington, D.C.

6. COASTAL ENVIRONMENTS AND OCEANS

Dr. Ellen Prager stands in the moon pool of the Aquarius undersea habitat in Key Largo, Florida, during a ten-day mission in 1998.

A CONVERSATION WITH
Ellen Prager

Having obtained her scuba certification in high school, Ellen Prager certainly spends her share of time literally immersed in the environment she studies, including a stint as host researcher for the Jason Project, in which she lives underwater for two weeks in the Aquarius undersea laboratory.

An experienced marine scientist and author, Prager has spent the last two decades conducting research and educating people about the oceans in areas as diverse as St. Croix, the Galapagos Islands, Papua New Guinea, and the Florida Reef. She has appeared on numerous television broadcasts and published several books including *The Oceans* (McGraw-Hill, 2000) and a children's book series with the National Geographic Society (*SAND*, 2000; *Volcanoes*, 2001, and *Earthquakes*, 2002).

Prager received her Ph.D. from the Coastal Studies Institute at Louisiana State University and M.S. from the University of Miami's Rosenstiel School of Marine and Atmospheric Science, where she was the former assistant dean. Now a writer and president of Earth$_2$Ocean, Inc., Prager's work focuses mainly on public education and outreach. "To be good custodians of the Earth, we must communicate effectively about its dynamic ocean-atmosphere system," she says. "Otherwise, we threaten the planet's health—and therefore, our own."

What Is the Issue?

Our oceans and coastal waters sustain life on Earth. Though vast and seemingly impenetrable, they supply billions in economic revenue and millions of jobs. They contribute to human health and play a critical role in security. But the ocean and its valuable resources are now at risk. They've been a dumping ground for decades, both intentionally and unintentionally. Years of misuse, poor management, and unlimited extraction have taken a toll.

From coastal tourism and recreation to trade at our docks and the seafood in our supermarkets, our economy and health are linked to the sea. Some estimate that in the United States alone, ocean-related activities generate over $800 billion each year in revenue, creating jobs on ships, in the navy, and at our docks, seaside motels, restaurants, and marinas. Much of our oil and gas comes from beneath the ocean floor, where we also extract sand and minerals for construction. And more than three billion people obtain a major part of the protein in their diet from seafood.

It's impossible to overestimate the importance of the ocean, as our very life support system depends on it. Oxygen is provided by its abundance of small drifting plants. Ocean currents balance the planet's heat. Through its great capacity to absorb carbon dioxide, the sea makes the Earth's atmosphere amenable to human life. The oceans also host an unrivaled wealth of organisms —a diversity and abundance even greater than that of our rainforests.

• Pollution

Discharges are easy to spot when they come from specific sources, like factories. Pollutants, however, also enter our ocean waters in more subtle ways, as unwanted hitchhikers on water flowing off the land, in rivers running to the sea, and on particles raining down from the sky. While we've made reasonably good progress over the past thirty years by identifying and targeting pollutants from single, or point, sources, we've just begun to tackle nonpoint source pollutants, which come from our cities and towns, cars and airplanes, and food production systems. The results are evident on our coasts. According to the National Resources Defense Council, in 2001 there were over 16,000 beach closings and swimming advisories issued across the nation due to water contamination. And the Environmental Protection Agency reports that 23 percent of the nation's estuarine area has been found impaired for swimming, fishing, or supporting aquatic life. According to the National Estuarine Eutrophication Assessment, many of the nation's bays and estuaries suffer significant oxygen depletion, sea grass loss, or toxic algae blooms—all symptoms indicative of nutrient excess, which is the result of increased pollution as nutrients enter the sea in runoff from sources such as agricultural lands, lawns, animal farms, and golf courses.

Excess nutrients are not the only serious problem in coastal and ocean waters. Oil, sewage, pesticides, contaminated sediments, marine debris, and the release of other chemicals and potential pathogens (such as bacteria and viruses) pose a significant threat. Human activities cause over 28 million gallons of oil to enter North American waters every year. The main problem is not large tankers or oil spills that grab headlines, but rather oil that comes from contaminated runoff and older two-stroke engines.

Air emissions also contribute to nonpoint source pollution. Particles and chemicals in the atmosphere combine with rain and fall into our oceans, rivers, and onto land, like the mercury now found in coastal waters and fish, for example. Agriculture, incineration, coal-fired power plants, industrial facilities, and motor vehicles are significant sources of atmospheric contaminants, including a large part of the new nitrogen entering estuaries.

Although we have greatly reduced pollution from sewage treatment plants and septic systems over the last several decades, Fleming, Katz, and Hammond state that more than 30,000 million liters of municipal sewage still enter the nation's coastal waters daily. Treatment removes some potentially harmful substances. However, we continue to release nutrients, organic material, human pathogens, pharmaceuticals, and toxic substances like metals and household or industrial chemicals.

• Coastal Population Growth and Land Use

Coastal populations are more than twice as dense as those elsewhere in the nation. Coastal development is often poorly managed, which helps to degrade marine environments. It also increases pressure on marine resources and leaves people vulnerable to coastal hazards, costing us millions of dollars. In the 1990s, southern California's population expanded by some 4,000 people per week and Florida's expanded by 4,400 new residents per week. Overall, coastal population growth is expected to continue at an astonishing average rate of 3,600 people per day, reaching 165 million by 2015 in the United States.

- Climate Change

Climate change is warming the seas, increasing sea level, and raising the potential for flooding, coastal erosion, coral bleaching, severe weather, and the spread of ocean-borne diseases. Changes in ocean temperature and sea level impact both human populations and marine life. Organisms unable to adapt rapidly enough to changing conditions in the sea may migrate to new habitats or die. Scientists believe that global warming is now a major factor in coral reef decline across the globe. Prolonged and high summer seawater temperatures in 1998, for example, caused significant coral bleaching, when corals lose their symbiotic algae due to extreme stress leading to coral mortalities worldwide. Warmer water may also make corals more susceptible to disease and storm damage.

- Overfishing

Fishing has been a way of life for generations in many communities, providing economic stability and a rich seafaring cultural heritage. However, because of dramatic fisheries declines and collapsed stocks, fishermen, their families, and entire communities are facing economic hardship. The ocean's ability to provide enough fish to meet future demand for seafood is in question. Diminishing fisheries are the result of over fishing, including the unintentional taking of nontarget species, or "bycatch," as well as too many fishing boats, pollution, and ineffective fisheries management. For more information, see chapter 12 on fisheries.

- Invasive Species

Worldwide, invasive species are living where nature never intended, and it's changing our coastal ecosystems and costing hundreds of millions of dollars annually. Most marine nonnative animals and plants are introduced through ships' ballast water—water carried to ensure stability and balance weight. The aquarium trade, floating marine debris, boating, fishing, and navigational buoys or drilling platforms also contribute to the spread of invasive species. In San Francisco Bay alone, there are more than 175 species of nonnative marine animals and plants. Asian and European shore crabs are now found along the coasts of New England and California, damaging valuable fisheries. The infamous zebra mussel has taken over the Great Lakes, clogging intakes for power plants and fouling hulls, pilings, and navigational buoys. And the list goes on.

- Habitat Destruction

We are losing critical and sensitive habitats such as wetlands, coral reefs, sea grass beds, kelp, and mangrove forests at an alarming rate. For instance, according to Thomas Dahl's *Wetland Losses in the United States 1780s to 1980s*, 91 percent of California's wetlands have been lost since the 1780s. The USGS reports that Louisiana is losing twenty-five to thirty-five square miles of wetland each year. As wetlands are lost, so too goes their capacity to maintain water quality, stabilize the shore, and provide refuge for wildlife. As we look at the problem more closely, we see that habitat loss is both a problem by itself and a result of the concerns noted above.

- Ineffective Policy, Science, and Education

An ineffective system of ocean governance compounds these problems. Today, more than sixty congressional committees manage nearly twenty federal agencies and permanent commissions with ocean-related activities. They are governed by more than 100 federal ocean-related laws. These laws and agencies are poorly coordinated, and there is a startling amount of redundancy within the federal system. Just as importantly, there is no one point of high-level leadership in the nation's ocean governance system. In fact, there is no comprehensive national ocean policy.

Investment in ocean science and technology has been stagnant and simply insufficient over the past two decades. It represents a fraction of the investment in space exploration. Less than 5 percent of the oceans have been explored, and we desperately need better information for more effective management. Furthermore, the general public's ocean literacy and stewardship is severely lacking. Resources, training, and experiential opportunities for teachers, students, and the public are insufficient. We lack necessary funding to implement effective, long-term education programs.

How Are Environmental Professionals Approaching the Issue?

To better manage the oceans, we must better manage the people who do the damage. This takes money, time, and political will. Box 6-1 outlines the strategies that environmental professionals are working on to improve ocean and coastal waters management, and the human activities that threaten their future.

• Generating Political Pressure

In 2000, the Oceans Act established a U.S. Commission on Ocean Policy to make recommendations for a coordinated and comprehensive national ocean policy. This had not occurred since the Stratton Commission in 1966, which resulted in significant steps toward improving ocean governance, including the creation of the National Oceanic and Atmospheric Administration (NOAA) as well as a series of important legislative acts such as the National Marine Sanctuaries Act, the Coastal Zone Protection Act, and the Magnuson-Stevens Fisheries Management and Conservation Act.

In early 2004, the U.S. Commission on Ocean Policy submitted its report to Congress and the President. It emphasized the need to build on the achievements of current federal programs and to provide high-level leadership in the form of an Assistant to the President for Ocean Policy and a National Oceans Council to improve coordination of agency efforts. Recommendations also suggest moving toward a more regional, precautionary, adaptive, and ecosystem-based management approach for marine issues.

The effectiveness of this commission and its report will be determined by whether national leaders take the bold actions and make necessary investment for significant change. Will they overcome partisan politics and political inertia, recognize the value of marine resources, and act on the critical need to make immediate change? As an optimist, I'm hopeful. As a realist, I'm skeptical. But one thing is certain: without committed **marine scientists** and

BOX 6-1

STRATEGIES FOR PRESERVING COASTAL AREAS AND OCEANS

- Generating Political Pressure
- Focusing on Prevention and Balance
- Taking a Regional, Ecosystem Approach
- Obtaining Adequate Information
- Educating Target Audiences

researchers to provide leaders with solid science, and activists committed to building public support for them, nothing will happen.

- Focusing on Prevention and Balance

While recognizing the utility and value of marine resources, today's experts are trying to better balance use with protection and sustainability. Rather than regulate or punish polluters only when crises arise, we must create policies that prevent pollution and provide incentives for environmentally friendly practices. Whether policies apply to fisheries, invasive species, or coastal development, this balance requires that policy makers, **marine scientists**, industry representatives, community members, and environmental advocates work together.

When there is agreement on particularly sensitive areas within a region, such as a wetland or nursery ground for fish, for example, experts can recommend they receive special protection. One method for ocean protection is to establish "no-take" marine reserves, where the extraction of all marine life and minerals is prohibited. Scientists such as Roberts, Bohnsack, Gell, Hawkins, and Goodridge writing in *Science* magazine have proven that within these areas, fish grow larger and are more abundant than in comparable unprotected areas. Experts can also work with community leaders and ocean users to identify areas where recreational boating, diving, and fishing are permissible, so that these activities don't disturb fully protected areas. When balance is achieved, implementing effective policies or establishing marine reserves can help make improvements occur relatively quickly, even within a few years. Unfortunately, many fishers and other users resist marine reserves, which is why very little of our coastal and ocean waters are fully protected.

- Taking a Regional, Ecosystem Approach

Ocean and coastal resources have generally been managed issue by issue, in isolation from one another and from outside influences. For instance, we have managed salmon without managing the fish they feed on or the integrity of the streams they breed in. Because of the inherent links within the ocean food web and overall environment, scientists now believe we must manage most fisheries on an ecosystem scale. This means they try to take into account the ways that multiple species interact, as well as the physical, chemical, biological, and geological environment of which they are a part. An example of this approach is in

147

MARINE SCIENTIST

AT A GLANCE

Employment:

21,000

Demand:

Good

Breakdown:

Academia, 35 percent

Public sector, 40 percent

Private sector, 20 percent

Nonprofit, 5 percent

Trends:

• Competition for jobs is fierce, as there are often more qualified candidates than jobs available.

• Marine and coastal issues are experiencing an increase in funding relative to other environmental issues.

Salary:

Starting salaries in marine science range from $29,000 to $36,000, with the median range from $48,000 to $54,000 annually. Top pay reaches $72,000 or more for scientists, and well over $90,000 for top tier professionals.

JOB DESCRIPTION

The fascinating work of marine science draws many people to this field because it offers opportunities to work with marine life on the edge of a new frontier in beautiful locations. Professionals working in this field include marine biologists, oceanographers, oceanic engineers, researchers, and college professors. Oceanographers in particular spend their time studying the evolution of the ocean bottom, its chemical and geological makeup, and tidal movements and currents. Oceanic engineers design and build equipment specifically used in marine science. All of these professionals can work at governmental agencies, nonprofit organizations, universities, consulting firms, and aquariums.

Marine scientists study saltwater organisms, investigate formations on the ocean floor, and study and minimize the impact of development on coastal areas. They also collect samples, and analyze sea and coral life and the effects of pollution. Frequently, they use technology such as remote sensing and Geographic Information Systems. At times, marine scientists' work occurs on a research ship, although scientists usually divide their time between fieldwork and analyzing data indoors.

Scott Donahue, a lead scientist at the Florida Keys National Marine Sanctuary, says his team restores coral reefs damaged by boats that run aground on them. "We try to come up with ways to restore the coral reef to a level equal to the way it was before it was damaged. This is vital because coral reefs, globally, are declining in health. We're trying to preserve as much as possible."

GETTING STARTED

• For entry-level employment, aspiring marine scientists should earn a bachelor's degree in fields like marine biology, geology, oceanography, and ocean engineering.

• A master's degree and Ph.D. will enable you to work in advanced research and academic positions.

• Look for jobs, volunteer opportunities, and summer programs that will introduce you to research methods. More career information is available at the Sea Grant website at www.marinecareers.net.

• Learning how to use Geographic Information Systems will give you an advantage.

• Getting your scuba diving certification will make you an attractive candidate to work in remote field sites, which should be expected in most jobs.

ADVICE FROM THE PROS

Dr. Judy McDowell, a professor and marine studies researcher at Woods Hole Oceanographic Institution, studies changes in habitats for fish and other marine life caused by environmental contaminants. "You need to constantly ask questions and solve puzzles to work in this field," she says. Instead of learning marine science as an undergraduate student, McDowell encourages students to gather a strong background in math, chemistry, physics, and biology. This will help ease students into graduate marine science studies.

the northern Gulf of Mexico, where regulators are trying to reduce fertilizer use along the Mississippi River watershed to improve conditions downstream for fish and other marine life.

Managing on an ecosystem basis, however, is not a simple task. Because ecosystems do not necessarily follow jurisdictional borders like state or county lines, they require a more regional management approach. This means that it's critical to have many stakeholders' inputs and a strong base of information and understanding for wise decision making. Part of this ecosystem approach must include planning and development that protects unimpaired, sensitive environments and manages plans for new growth in already impacted areas. To do this, **elected officials**, local resource managers, local water users, and those within industry and the scientific community must assess regional resources and identify critical habitats as a basis for planning and decision making. Then, of course, people with very different interests must agree on top priorities and develop realistic action plans that they can feasibly implement. One example where people have come together to manage on an ecosystem basis is the Chesapeake Bay region, where the Environmental Protection Agency, the states of Maryland, Virginia, and Pennsylvania, and the District of Columbia have partnered with scientists, private groups, landowners, conservation groups, and citizens to protect water quality and living resources in the watershed area. In particular, these groups work together to preserve the Chesapeake Bay estuarine system by working on nutrient reduction, toxics management, sustainable development, and habitat restoration programs that aim to prevent environmental degradation and maintain the ecosystem's long-term health.

- Obtaining Adequate Information

Managing ocean and coastal resources should be based on the best available science. In fisheries, for instance, if we understand where fish spawn and what environmental conditions they need to survive, grow, and reproduce, we can identify and protect critical spawning and nursery areas.

But often, either the science is lacking or it's not in a format that decision makers can use. Environmental professionals from a broad range of disciplines are now attempting to identify and fill in scientific knowledge gaps and make their results more accessible.

Creating an integrated ocean observing system is an important step toward providing much needed valuable information. This system will serve decision makers and **marine scientists** by providing timely ocean observations, much like the system we use to forecast and study weather. An observing system is already at work in the tropical Pacific Ocean. It combines observations on buoys, ships, and from satellites with computer modeling techniques to predict the development of El Niños. The forecasts can now be used to issue warnings to fishermen, farmers, weather forecasters, and others whose livelihood is influenced by ocean and/or climate conditions.

Because ocean observing systems require myriad measurement techniques, hundreds of scientists, marine technicians, and agency representatives with requisite skills will be in demand. The development and maintenance of a national observation infrastructure will be a major source of marine science employment over the next ten years and beyond.

• Educating Target Audiences

Everyone from policy makers to resource managers, teachers, and individuals should understand the importance of the oceans, their relevance to everyday life, and the challenges we now face in long-term restoration and protection. For this reason, education is essential for improving coastal water and ocean conditions and preserving marine resources for the future. Effective education is accomplished through a variety of means, including formal and informal efforts in classrooms, aquariums, or museums; over the Internet; and at local and regional sites.

Many people are currently working to improve ocean education. The Center for Ocean Science Education Excellence (COSEE), for instance, fosters coordination of ocean education resources. The Consortium for Oceanographic Research and Education (CORE) runs the National Ocean Science Bowl, a wonderfully inspiring national academic competition for high school students. The Virginia Institute of Marine Science's Bridge project offers a quality-controlled, web-based ocean education clearinghouse. And the Jason Project, a multidisciplinary education program that enables students and teachers to do field work from the classroom, brings expeditionary science into middle schools. As a result of these and other educational efforts, oceanographers, marine ecologists, coastal zone managers, **conservation biologists**, and coral reef scientists are helping to train the next generation of marine professionals.

PROGRAM ASSISTANT

AT A GLANCE

Employment:

32,000

Demand:

Excellent

Breakdown:

Nonprofit, 55 percent

Government, 35 percent

Private sector, 10 percent

Trends:

• As organizations seek to do more with less, program assistants are asked to take on some responsibilities of the managers they serve and/or the clerical duties that would have previously gone to administrative assistants. The pressure to move up and down the ladder of responsibility makes this a stressful job.

• Program assistants are often thought of as managers-in-training, to be promoted if funding allows.

Salary:

Entry-level salaries range from $25,000 to $33,000, with the median annual income around $36,000 and $36,000. Assistant at well-funding organizations earn $44,000 or more.

JOB DESCRIPTION

Behind every successful environmental project, program, or campaign is at least one program assistant making sure everything runs smoothly.

Program assistants are the primary contact point for those seeking information from a nonprofit organization or government agency, so it's essential that they know all details of a project or program. They often work on many projects, quickly getting a glimpse of nearly everything that's involved. They may coordinate fundraising campaigns, research new donors, write press releases, arrange media interviews, or design organization newsletters and websites to ensure that the public receives accurate information.

Adrienne Wojciechowski, program analyst for The Nature Conservancy's governmental relations department in Washington, D.C., focuses her efforts on forestry and agriculture programs. In each program, Wojciechowski coordinates a team of policy organizers and professionals who advocate for woodlands or farmlands preservation. She also plans fundraisers to help The Nature Conservancy acquire lands, such as the recent protection awarded to 3,600 acres in Georgia. "The ultimate challenge for us is to preserve the last great places in this country," she says.

Program assistants also provide administrative support for their managers by organizing meetings, receptions, and conference calls; coordinating staff schedules; arranging travel plans; and completing expense reports. They often draft correspondence and proofread, edit, and type legal, technical, scientific, and public education documents. They file, copy, mail, and fax items as needed. All of these responsibilities mean that program assistants must be well organized, detail-oriented, and multiple-task-oriented professionals able to operate in a fast-paced environment.

GETTING STARTED

• While most positions require a bachelor's degree, no specific college major is needed for employment. However, knowledge of environmental fields and issues is helpful.

• Interning or volunteering at environmental and social justice organizations will provide valuable experience.

• Experience in fundraising, either in a development office or at a nonprofit organization, will make you more marketable.

• Website design and management is useful for this job.

ADVICE FROM THE PROS

Gretchen Muller, project manager for the National Audubon Society's bureau in Washington, D.C., began her career with the society as a program assistant. In her current and former position, she's responsible for overseeing projects about wetlands protection and endangered species. "As you're doing this work, it's very important to be flexible and willing to take on many tasks," she says. "You have to be able to do whatever needs to be done, keeping in mind that it's all for the greater good." Muller advises that aspiring program assistants find an internship on Capitol Hill to learn how politics and government operate. "You'll have a better chance getting a job if you know what's going on."

Career Spotlight

The Ocean Conservancy
Washington, D.C.

When Margaret Mead said that "a small group of committed citizens can change the world," she could have been describing the staff of the Ocean Conservancy—arguably this nation's foremost advocate for the oceans. Headquartered in Washington, D.C., with offices in California, Virginia, Florida, Maine, Alaska, and the Virgin Islands, The Ocean Conservancy (TOC) employs only eighty people to pursue an aggressive international agenda of protection for marine fish and wildlife, coastal waters, and ocean ecosystems. Although small, the conservancy's band of scientists, lawyers, and lobbyists seems to be involved in every issue affecting the sea.

With influential work and limited jobs, it comes as no surprise that TOC attracts remarkable people—like Sierra Weaver. Weaver was hired in 2003 as program counsel for the Marine Wildlife Program. A 2001 graduate of Harvard Law School (where she was managing editor of the *Harvard Environmental Law Review*), Weaver worked for two years as Dockery Environmental Law Fellow at the Southern Environmental Law Center, where she honed her litigation skills on air pollution cases in Atlanta and wetlands protection of the North Carolina coast.

"Protecting whales, dolphins, otters, and manatees is my job," Weaver says. And she uses every tool she can to do it. Since coming to TOC, she has negotiated with fisheries managers and Hawaii fishermen on behalf of endangered sea turtles; helped the team lobbying for reauthorization of the Marine Mammal Protection Act; advised TOC's Florida office on manatee protection; and worked on myriad other litigation, legislation, and administrative projects, all while working out of TOC's Washington, D.C., headquarters.

Far from the intrigues of Washington, Nick Drayton walks the beaches of St. John in the U.S. Virgin Islands. As TOC's Caribbean ecosystems program manager, Drayton focuses his attention on advancing the establishment of "Marine Protected Areas" in the Virgin Islands as an effective tool for protecting the area's spectacular—and fragile—coral reefs. Eschewing the desk-bound life, Drayton works closely with local and federal agencies, other non-governmental organizations (NGOs), and local people. He scuba dives with area fishermen to identify fish spawning aggregations that depend on

Work is just a day at the beach in the Virgin Islands for Ocean Conservancy Manager Nick Drayton.

the reefs and has secured an appointment as the coral reef "local action strategy" coordinator by the territorial government.

Drayton's master's degree in coastal biology from the University of North Carolina/Wilmington is complemented by years of experience as the regional coordinator for the Caribbean Conservation Association, director of the British Virgin Islands National Parks Trust, and The Nature Conservancy's director of conservation programs in the Eastern Caribbean. Like so many other successful NGO directors, Nick started his career as a volunteer.

Whether speaking out for sea turtles or protecting coral reefs, success depends on public support—and it's media attention that makes the difference. Tom McCann is the organization's media relations manager, charged with national and regional campaigns. He pitches story ideas to newspapers, magazines, websites, radio, and television; teaches TOC staff how the media works; and monitors opportunities and trends. "People love the oceans, but it can be tough to maintain media interest in endangered smalltooth sawfish, a National Marine Sanctuary for the Channel Islands, or rebuilding groundfish population in New England. My job is to find the narrative that will turn on editors who are swamped with material."

After securing a B.A. in environmental studies and psychology from Wheaton College, McCann gained experience in both the private and nonprofit sectors as an account executive with the environmental practice at Porter Novelli, as director of public relations for World Resources Institute, and as an independent communications consultant.

TOC promotes "science-based" advocacy, and it delivers by hiring people like Dr. Cheri Recchia, director of the Ecosystem Protection Program. Recchia studied beluga whales on her way to a Ph.D. in biological oceanography from MIT and Woods Hole Oceanographic Institution. She then led projects at World Wildlife Fund Canada and for the Great Barrier Reef Marine Park Authority in Australia.

A superb example of TOC's global challenge is Recchia's current effort to promote areas that should be protected for many uses in the San Andres Archipelago of Colombia. "We're working with the government there to craft a plan that will protect biodiversity and promote community stewardship, and protect traditional uses by the native community," she reports. "This is engaged science, and it's the best hope for lasting protection of the ocean."

Career Advice from the Expert

Many people dream about a life connected to the power and beauty of the sea. Marine science is an exceptionally rewarding career. For some, it verges on obsession. However, marine science does not necessarily mean becoming a research scientist or teaching in academia. Aspiring ocean professionals can also become marine technicians, divers, graphic or computer specialists, policy makers, educators, or even professors who work with nonprofit conservation organizations.

Each career track requires a slightly different set of skills, experience, and academic background. However, it is wise to take basic courses in biology, chemistry, geology, math, earth systems, and physics. These can then be complimented with courses in oceanography, marine science, and more advanced classes in graduate school. Getting relevant experience is highly recommended, even when the pay is minimal. I got my scuba certification in high school and worked as a safety diver for an undersea laboratory one summer during college. I got the job by having the necessary skills and simply asking for it. Unless you are willing to risk rejection, the best opportunities may never come your way.

No matter what your area of interest, don't be afraid to try something, even if it isn't exactly what you want. Along with discovering what you want to do, you must also determine what you don't want to do. I once worked for Shell Oil in a downtown skyscraper in New Orleans. After a long six months, I decided it just wasn't for me and applied for a fellowship and Ph.D. program. I don't regret working in the corporate world, because I gained great experience and insight, but will never again wonder if it could have been a career track for me.

Prospective professionals should begin learning about opportunities by working and networking within the ocean community through volunteer and internship programs such as those offered by aquariums, parks, education centers, state coastal zone management programs, NOAA, USGS, and respected labs like Scripps, the Rosenstiel School, and Woods Hole. The Sea Grant institutions, a national network of universities and research institutions that encourage stewardship of our marine resources through research, education, outreach, and technology transfer, are also places to gain valuable experience. The Internet is a great resource to search for opportunities and contacts.

With the growing necessity for improving ocean and coastal management and education, developing policy based on good science, and creating a worldwide ocean observing system, there should be many future employment opportunities in this field. This is an exciting and imperative time to dive in and get your feet wet, literally!

RESOURCES

Chesapeake Bay Program, www.chesapeakebay.net

Consortium for Oceanographic Research and Education, www.coreocean.org

Monterey Bay Aquarium, www.mbayaq.org

National Oceanic and Atmospheric Administration, www.noaa.gov

National Sea Grant Program, www.nsgo.seagrant.org

Office of Naval Research, www.onr.navy.mil

Rosenstiel School of Marine and Atmospheric Science, www.rsmas.miami.edu

The Bridge, www.vims.edu/bridge

The Ocean Conservancy, www.oceanconservancy.org

U.S. Geological Survey, www.usgs.gov

Woods Hole Oceanographic Institution, www.whoi.edu

7. ECONOMICS

Michel Gelobter, executive director of Redefining Progress, speaks about economic policy and justice.

A CONVERSATION WITH
Michel Gelobter

At the beginning of the twenty-first century, more than ever before, people all over the world are questioning the global economy and its contributions to an unsustainable society. Although economists and citizens alike have challenged our current economic institutions, very few have come up with answers that will promote and achieve sustainable economic change that will result in an improved quality of life for all people worldwide. Instead, they continue to grapple with the question of how we go from recognizing a problem to creating change that will benefit people, the environment, and the economy. Michel Gelobter believes that "To change behavior, we need to change systems, which we must do by changing ourselves."

Gelobter's ambition is evident. He is currently executive director of Redefining Progress in Oakland, California, a nonprofit organization that creates sustainability by measuring the real state of the economy, designing public policies to change undesirable behavior, and implementing new economic frameworks to create an alternative economy.

Before coming to Redefining Progress, Gelobter founded the Environmental Policy Program and Columbia's University's School of International and Public Affairs. He also served as director of environmental quality for the City of New York. He obtained his Ph.D. from the University of California-Berkeley's Energy and Resources Group and has written broadly about environmental justice, global warming, sustainability, and commons management.

What Is the Issue?

What we count, counts. Fundamentally, economists count the value of things in dollar terms—in prices. Prices not only help us to buy things in the grocery store, but they also help us to decide as a society whether one policy or another will benefit us more. One reason we have environmental and social crises is because, as a society, we count the wrong things.

Consider the Gross Domestic Product (GDP), which is the economic measurement by which we measure our economic (and often social) well-being. The GDP is a misleading standard because it counts all economic transactions as positive contributions to our economy, culture, and quality of life. If a person has a heart attack and spends money in the hospital, or goes to jail instead of college, the money spent grows our GDP simply because money changes hands.

Environmental degradation is very high on the list of things the GDP doesn't measure. Many products on which we spend money depend on the environment as a free place to dump pollution. The real cost to the environment is not reflected in prices we pay.

From a sustainability perspective, another deep accounting flaw relates to what is counted as income versus wealth. When we cut down a forest for wood products, the revenue received from selling lumber and paper is counted as income in the year they are sold. In reality, this wood represents the liquidation of our natural capital grown over a period ranging from thirty to hundreds of years. In cashing that capital out, we are mistakenly treating it as a one-time cash infusion instead of being depreciated over the many years required for replacing it from Mother Earth. Finally, leaving a healthy forest ecosystem alone doesn't generate any GDP. Cleaner air and water, space for biodiversity, and beautiful vistas derived from a stand of trees never makes it into our accounting books.

We can see other examples of such accounting errors throughout our economy. Pumping oil, laying off workers, draining aquifers, funneling youth to prisons, mining coal—all these activities extract natural and social resources that took years to develop, but only minutes to waste forever. Yet we label them as growth, even though true accounting would show massive deficits instead of rampant growth.

In addition to reducing "growth" to purely financial accounting, the GDP discounts many positive activities that people universally agree are important.

When parents stay home from work to care for their children, or when citizens volunteer to plant trees, build literacy, or restore a wetland, the GDP is unmoved. Because these essential community-building acts do not involve economic transactions, they are not counted.

- No Such Thing As a Free Lunch

We rarely incorporate environmental damage as a cost of doing business. But there's no such thing as a "free lunch." For example, when we drive cars, we pay for the gasoline, tires, oil, and car itself; but we don't pay for tailpipe emissions and dust from roads, which turn into particulate matter and threaten people's health. Instead, these negative impacts, which economists call externalities because the firms that create them don't bear (or internalize) the costs, show up as human health and productivity impacts that add to the GDP. In Los Angeles alone, air pollution costs upwards of $15 billion per year in excess death, missed workdays, reduced productivity, and restricted children's activities due to asthma and other respiratory diseases. If we were to internalize the true cost of driving a car in Los Angeles, a different economy would emerge that valued mass transit, dense inner cities, and reduced sprawl. Overall, an economy that conscientiously accounts for environmental costs, and who in society bears those costs, also grows new options for buying, consuming, and living more sustainably.

- Unhealthy Subsidies for Unsustainable Behavior

Our economy not only ignores externalities, but also directly subsidizes some of the most egregious environmental abuses. Within the United States, we massively subsidize logging, energy exploration and extraction, and a host of unsustainable agricultural practices. We underwrite the worst forms of transit, airports and freeways, while starving far more efficient mass transit and railroads. Incentives like these increase pollution and ecological degradation to the detriment of all.

- Inequality and Equity Economics

The current economic system creates and perpetuates inequality on many levels, partially because so many economists focus on what Dr. Seuss' Lorax called "biggering." It's far easier to calculate how to maximize profits than it is to calculate how to maximize fairness. Amartya Sen won the Nobel Prize in Economic

Science in 1998 for a career dedicated to formalizing the economics of equity and social justice. He asked, and answered, questions about the global distribution of wealth, growing poverty, and people's inability to lead fulfilling lives because they lack basic entitlements like food, water, shelter, and democracy. He formalized the concept of relative inequality, which explains how, even though a poor person in the United States has ten times the cash wealth of a poor person in India, he can be worse off in the most important ways—in his ability to control his life and to live to his rightful potential.

When we understand economic growth narrowly by counting all expenditures as progress, we neglect important equity dimensions. A dollar spent fueling a Jaguar actually contributes less to well-being than a dollar spent feeding a hungry child or immunizing a sick person.

The wealth of any nation depends on common assets—the natural, economic, and social resources that initially are public goods belonging to all people. When those assets become the private monopolies of oil and timber companies, mining interests, real estate speculators, and software companies, the public is seldom compensated. Similarly, when publishers and software companies gain unreasonable control over socially produced ideas, images, and information, the public loses part of its cultural heritage. As a result of this "enclosure" or privatization, the gap between rich and poor has widened and the public domain has shrunk. Additionally, the philosophies that justify privatizing commonly held resources rationalize assaults on government-produced public assets like health systems, educational institutions, libraries, parks, pollution control, and climate management, by claiming that the public good is just the sum of private wealth.

• Threats to Natural Wealth

To combat threats to the earth's climate, oceans, fisheries, or forests, we must reclaim these resources as common assets. Unfortunately, the privatization of these common resources is accelerating. In this way, they have much in common with many "new economy" goods, like the Internet, broadcast spectrum, or genetic information. But common assets do not naturally belong to any private company or individual. For example, recent legislation may provide fishers and fishing companies with rights to trade fishing permits that are based on quota systems determining total allowable catch. In theory, tradable permits will eliminate overcapitalization and increase efficiency; however, by defining natural wealth in a bald market frame, tradable fishing rights could enable a handful of companies to buy rights to almost the entire U.S. catch. The next step would be monopolies controlling all fish catching, processing, and marketing. Irresponsible privatization of common assets, like fish, will accelerate their depletion. We must incorporate stronger protections for the commons, even if the private sector primarily manages them.

• Globalizing Our Economic Growth Models

Even with all of the problems already mentioned, we often tout our "free-market system" as superior to that of "less developed" countries, particularly with respect to environmental protection. But the western model of economic growth is often disastrous in pervasive ways for both global and local environmental concerns. The United States alone accounts for 35 percent of total greenhouse gases in the atmosphere today, 23 percent of which can be attributed to carbon dioxide emissions from car traffic alone. As our "developed" standard of living increases, these problems intensify. More Americans are driving sport utility vehicles, and, according to the Sierra Club, they put out 43 percent more global warming pollutants and 47 percent more air pollutants than the average car.

Globalization continually promotes western and American transportation and overall lifestyles, including car-reliant cities and sprawl. Unless we alter behaviors in the west, we cannot expect the rest of the planet to follow suit in curbing global resource consumption. Key arenas for this struggle are institutions like the World Bank and the International Monetary Fund. Well beyond their purview of loan and debt repayment programs that often fail and contribute to poverty and desperation, these organizations and others must grapple with the environmental and social dimensions of the economic models they promote.

How Are Environmental Professionals Approaching the Issue?

• Creating Alternative Accounting Measures

Amartya Sen's Nobel Prize represents growing recognition of how the economics of raw growth does not adequately respond to the challenges of our times. The movement for creating alternative economic and ecological indicators is growing. These indicators are necessary benchmarks for determining

BOX 7-1

CHANGING THE CURRENT ECONOMIC SYSTEM

- Creating Alternative Accounting Measures
- Reforming Taxes and Diverting Subsidies
- Implementing the Precautionary Principle
- Putting Value on Common Assets
- Rethinking Global Institutions

what we're really doing to our environment and how to shift our economy toward sustainability.

The "Genuine Progress Indicator" (GPI) is one alternative to the GDP. In calculating economic growth, the GPI adjusts the GDP for income inequality, environmental degradation, and other unsustainable behavior. It adds both positive and negative values that are not adequately included in current economic measures, including time people spend with their families, volunteerism, environmental impacts of air pollution, commuting, and equity of incomes in a specific area. Placing value on the environment and society allows for more accurate reflections of economic growth. Our calculations show that while the GDP has gone up 295 percent since 1950 and stood at $32,709 per capita in 2002, the GPI only rose 168 percent and stood at $9,426 during that same time.

The "Ecological Footprint Index" is another alternative measure that directly accounts for the earth's natural capital. It simply adds up the land necessary for our current production and consumption patterns, including the land necessary to grow wood in forests, raise animals for food, construct buildings and infrastructure, and sequester the carbon we emit in burning fossil fuels. By comparing accounts of the regenerative capacity of key resources to humanity's consumption, the Index shows dramatically that we are using approximately 20 percent more planet than is available, and many natural resources are in danger of, or already have, collapsed.

No single measure can cover every economic situation. For this reason, community-driven indicators are critical in defining and protecting what is

EXECUTIVE DIRECTOR

AT A GLANCE

Employment:

31,000

Demand:

Excellent

Breakdown:

Nonprofit sector, 100 percent

Trends:

- Increasingly, organizations are hiring executive directors from nonenvironmental fields if they are skilled leaders with organizational development abilities.
- Many executive directors get their start in entry-level positions at nonprofit and advocacy groups, and work their way up the ranks over time.

Salary:

Entry-level positions range from $30,000 to $65,000, with median salaries around $80,000. Top salaries regularly exceed $100,000, and the biggest groups can pay $200,000 or more.

JOB DESCRIPTION

Nonprofit environmental groups are often able to influence meaningful policy changes due largely to the leadership of their executive directors. For those with significant nonprofit experience, working as an executive director of a nonprofit group affords a chance to champion a cause with others, while developing and enhancing the mission of the group. Accordingly, Tom Mooers, executive director of Sierra Watch in California, says that much of his time is spent "working with the board and staff to establish and implement a long-term vision for the organization."

Regardless of an organization's size or objective, executive directors usually have experience working for environmental causes and must be committed to change, as they are the frontline representative of the group. Executive directors must also be leaders who have strong fundraising abilities and networking skills. As such, grant writing, effective solicitation, and fundraising planning are essential skills for people in this position. "For any job in the environmental field, you can have good ideas for the cause, but you must raise money to support that cause," says Greg Small, executive director of the Washington Toxics Coalition in Seattle, a group that attempts to prevent pollution problems in Washington State. "Without funding, good ideas die a quiet death."

To ensure a nonprofit's continued success, executive directors spend a great deal of time meeting with other nonprofits, governmental agencies, local and corporate businesses, and the public to build relationships and foster collaboration.

Though some executive directors are the only employees in an organization, most are responsible for overseeing at least some staff and volunteers. For this reason, they must manage and motivate their staff to meet the goals of the organization. On a purely managerial level, they also ensure that daily operations run smoothly so as not to disrupt mission work.

GETTING STARTED

• A bachelor's degree is the minimum requirement, though most executive directors hold a master's degree.

• Look for fundraising experience to make yourself more marketable.

• Volunteer for student advocacy groups to gain grassroots experience and an advantage in this competitive field.

• Jobs or internships where you are responsible for managing other staff members will provide valuable experience.

ADVICE FROM THE PROS

"The solution for a lot of these problems is, in large part, tied to the need for a social change movement," says Greg Small. "The challenges we face are enormous opposition from powerful groups with a lot of money behind them." Small says that, to work as an executive director, people need to recognize what they're up against. "This means you have to have patience and perspective. You can make enormous change, it just takes time."

important economically, environmentally, and socially in specific locales. Community indicators are measurements defined through community visioning and constant assessment of desired conditions. The most effective indicators rely on participatory processes and actually drive changes toward a more sustainable community.

This model, pioneered primarily in progressive municipalities around the world, has helped citizens collectively improve environmental quality, economic vitality, and social equity in their communities. The city of Santa Monica, for example, has developed community indicators to measure conservation, transportation, pollution prevention, public health protection, and community and economic development. The city's sustainability coordinator and municipal employees of the Environmental Programs Division devised statistically measurable indicators and targets in each of these categories. The program has helped drive sustainable community changes, such as reducing citywide water usage by 6.3 percent between 1990 and 2000; increasing the number of city fleet vehicles that use reduced emissions fuels from 10 percent in 1993 to 70 percent in 2000; reducing city sewage flows by 14 percent between 1990 and 2000; and increasing the number of affordable housing units by 47 percent between 1990 and 1998. The city is also pursuing new targets in each of these categories to achieve for 2010.

- ## Reforming Taxes and Diverting Subsidies

Knowing where we stand is critical for achieving environmental sustainability. Changing the economic and fiscal policy drivers is the next step. As mentioned before, the U.S. tax code subsidizes environmental destruction far more than it promotes ecological and human health. This problem extends beyond the environment to social issues, including the employment rate and social security. If we could shift to efficient pricing structures or impose taxes for harmful environmental behavior, we could significantly lighten the burden borne by employers and employees alike.

While environmental charges do exist, they are not always effective. For instance, the United States does not tax fossil fuels in ways that directly wean us from unhealthy dependence on foreign oil. Charging the right price for resource use would keep tens of billions of dollars in the U.S. economy that is otherwise leaving. More significantly, it would boost productivity in almost every economic sector. For example, one form of a proposed carbon emissions tax could generate between $80 billion and $300 billion per year in new revenues while actually growing a stronger U.S. economy.

Economists at Redefining Progress are also working to internalize transportation costs to create similar benefits. We've evaluated the true costs of driving —from national security to the time people waste commuting—to highlight subsidies that allow for increased traffic and time spent on the road, rather than increasing mass transit use and decreasing the distance between where people live and where they work, play, or pray. Implementing congestion pricing can make people pay for their driving based on use and drive us to more sustainable solutions.

In order to be effective, tax reforms and new policies must remove subsidies that promote environmentally harmful practices. Costs will then better reflect the environmental, social, and economic values of products and services. A more environmentally friendly economy will also allow capital investment to flow away from ecologically destructive facilities, such as oil refineries, for example, to economically and socially productive activities, such as information products and urban infrastructure. This approach can deliver a "double dividend" for greater economic and environmental health.

- Implementing the Precautionary Principle

Traditional cost-benefit analysis does not take uncertainty into account. Most environmental problems, like climate change or toxic releases in water, however, are characterized first and foremost by the uncertainty of their impacts on health and ecological systems. For this reason, many people around the world use the "Precautionary Principle" to integrate the risks of industrial activities into an estimation of their benefits.

Toxicologists, for example, have found numerous examples of toxic products, from pesticides to pharmaceuticals, that have been more costly to society in the long run than their short-term commercial benefits. Using the Precautionary Principle in this way can help U.S. communities and policy makers shift the burden of proof away from polluters and toward a healthy and precautionary approach that will allow industrial designers to safely introduce new technologies into the world. Implementing the Precautionary Principle will help designers, researchers, economists, and legislators move us toward environmental policies that restore the earth, rather than simply cleaning up messes.

- Putting Value on Common Assets

At the turn of the twentieth century, wealth was accumulated from material resources and infrastructure carved from largely unbound land. Through homesteader and railroad grants, mineral leases, and unregulated use of air, water, and land, the United States grew to be the great industrial power of the world.

The wealth of the twenty-first century will be seen in a similar light, in that wealth will be built from exploiting riches belonging to all citizens. Genetic structures of life, the Internet, bandwidths that underlie all telecommunications, the regenerative capacity of the planet, and even how we divide our time between work, sleep, and play have all created formidable markets. Most would agree that we have only just begun to tap their potential value.

The Sky Trust is an initiative designed to counteract the exploitation of such public goods. It puts a particular value on "the sky," or atmosphere, as it relates to greenhouse gas emissions and encourages all citizens to exercise their right to clean air. Because the sky is a public good, this initiative argues that every child at birth is endowed his or her own share. In order to promote cleaner air for the future, it suggests that the government tax carbon emissions from fossil-fuel industries to then force private businesses to recycle money back to children, perhaps in the form of a savings bond that could encourage education. This proposal, although particularly aimed at reducing climate change, lends a lesson to any initiative looking to create an understanding of common assets, because it seeks to create an equitable and improved quality of life for all people, rather than an improved economy for industry and corporations.

- Rethinking Global Institutions

Creating new international institutions that hold the United States and other countries and their economies accountable for global social and environmental conditions is critical for our planet's health. New economic institutions must embody fair competition; address widening disparities in wealth, health, and well-being; and dig out from under a half-century of debt-driven development.

Creating an economics of sustainability will fundamentally alter the terms of trade between countries. Just as sustainable economic indicators will influence national policies, they will also affect currencies and international exchange. For example, assume one country has internalized environmental

ENVIRONMENTAL ECONOMIST

AT A GLANCE

Employment:

1,000 positions in 2004

Demand:

Rising

Breakdown:

Public sector, 50 percent

Private sector, 50 percent

Trends:

• Environmental economists will be needed for defining economic benefits of environmental alternatives, such as renewable energy sources, as they become more accessible to the public.

• An increasingly competitive economy and private companies' desire to expand into new markets will increase demand for environmental economists to guide adherence to government regulations.

Salary:

Starting salaries range from $35,000 to $50,000, with median incomes reaching up to $70,000 a year. Top salaries reach to $100,000 or more.

JOB DESCRIPTION

As development progresses at a remarkable rate, so does the need to incorporate environmentally sound practices into future economic policy. "We're here to promote an understanding of how environmental regulations affect the nation's economy," says Brett Snyder, environmental economist for the National Center for Environmental Economics at the U.S. Environmental Protection Agency. "We also help articulate the economic benefits of environmentally sound practices."

Unlike traditional economists, environmental economists take into account the environmental impacts of economic policy. Richard Newell, an environmental economist working at Resources for the Future (www.rfe.org), says that many of the world's environmental problems are based in economic consumption and production patterns. Environmental economists try to change these patterns by developing and advocating economic policies that incorporate the need to conserve natural resources and have as little impact on the environment as possible. As a result, these policies often highlight the benefit of environmentally friendly business practices.

Environmental economists spend much of their time conducting research at the local, global, or national level to assess the benefits and drawbacks of new projects or legislation. This work often involves collecting, analyzing, and comparing data—such as comparing the costs of gas and renewable energy sources—and sometimes conducting interviews or surveys. Environmental economists often collaborate with social scientists, government officials, and natural scientists to develop economic models with an environmental bent to encourage conservation and advise government authorities. In addition to working in government, environmental economists also work at research and policy organizations, consulting firms, international institutions like the World Bank and the International Monetary Fund, or in academia.

GETTING STARTED

• A bachelor's degree in economics is the minimum requirement for employment, although a master's or doctorate degree is preferred.

• To work in advanced research positions, a master's degree or, more often, a doctorate in economics, ecology, or natural resource management is necessary.

• Volunteer or obtain an entry-level job at a research firm or government agency to gain valuable experience.

• The Association of Environmental and Resource Economists (www.aere.org), a trade organization in Washington, D.C., provides valuable networking services.

ADVICE FROM THE PROS

David Austin, environmental economist for the Congressional Budget Office in Washington, D.C., just completed a research project on automobile fuel economy. His project involved searching for ways to conserve gas resources without raising gas prices too much. "I'm passionate about not wasting natural resources," Austin says. "And I think you must have that drive to work in this field." He suggests pursuing an internship with an environmental think tank while pursuing a Ph.D. "There's no doubt that having a doctorate is a real advantage," he says. "But I suggest getting a taste for the research life first because it may not be for you."

Career Spotlight

The Gund Institute for Ecological
Economics at the University of
Vermont
Burlington, Vermont

Although we can imagine a day when "ecological economics" is common work in business and government, that day is not yet here. For now, a great deal of this work happens in the academic world. Among the handful of leadership institutions in the field, the Gund Institute for Ecological Economics is consistently ranked at or near the top.

Located at the University of Vermont in Burlington, the institute has a full-time staff of seventeen people, including eight research professors, six research associates and assistants, two affiliated scientists, and a single program coordinator. This core staff is complemented by a cadre of shifting research fellows currently numbering twenty-nine students and scholars from a wide variety of different fields.

"Our task is to integrate the study and management of nature's household with the household of human beings," says institute founder and director Robert Costanza. "People traditionally call the first 'ecology' and the second 'economics,' and then set up artificial boundaries to separate them. We need to transcend those boundaries in our work because they just don't exist out there in the real world."

Given his background, Costanza speaks with some authority. Besides founding the institute, Costanza co-founded the International Society of Ecological Economics and was the chief editor of the journal *Ecological Economics* from its inception until late 2002. His eclectic academic background seems like odd preparation for a career in environmental economics—a master's degree in architecture and urban and regional planning from the University of Florida followed by a Ph.D. in "systems ecology" from the same university, with a minor in economics.

Traditionally trained economists, it turns out, are rare among the institute's faculty, staff, and fellows. Consider Matthew Wilson. Wilson is working on an ecosystem services valuation analysis for the state of Maryland—an effort aimed at creating new tools to help leaders incorporate the economic value of healthy natural systems into decision making about development in that already crowded state.

He laughs when asked about a "typical" career path. Wilson started out with a B.A. in anthropology from Colorado College, then moved on to an M.S. in geography from the University of Colorado at Boulder, and finally earned a Ph.D. in sociology from the University of Wisconsin (Madison), with a minor in regional planning.

"On this staff, we have ecosystems ecologists, urban planners, agricultural economists, environmental educators, mechanical engineers, policy analysts, foresters, marine scientists, international relations specialists, GIS specialists, mathematicians," he reports. "And we need them all."

That combination of skills and backgrounds is the strength of the institute, according to Jennifer Jenkins. "It's been said that society has issues, and a college has disciplines," Jenkins says. "That's not the case here. We let the problems themselves determine the mix of perspectives needed to address it."

For her part, Jenkins works on global climate change, particularly studying the role that forests play in the carbon cycle and might possibly serve in sequestering carbon dioxide to reduce its accumulation in the atmosphere. "My projects require an understanding of monitoring and inventory procedures, remote sensing, ecosystem modeling, as well as basic forest science," she says. As preparation for this work, she has a master's

degree in forest science from Yale and a Ph.D. in ecosystem ecology from the University of New Hampshire.

As the institute's work becomes more international, it requires graduate research assistants like Erica Gaddis. Since earning an M.S. in environmental science/policy from Central European University in Hungary, Gaddis has examined innovative ecological approaches to water remediation and treatment in Europe, Ukraine, China, Hawaii, and Maryland.

"Recently, my work has turned toward identifying biological mechanisms for removing low concentrations of endocrine disruptors and estrogens from water," Gaddis reports. "Like everything we work on here, it's an issue that requires us to combine ecology, public health, economics, technology, and governance. For me, that's the very definition of ecological economics."

As he looks ahead, institute director Costanza is bullish on the future. "Integrating ecology and economy is the great challenge of the twenty-first century. Professionals who can tackle this problem will be in great demand."

costs, and measures growth using the GPI. Should it have to compete globally with another country that rampantly destroys its resources and does not account for the costs it extracts from natural capital? Reforming national economic and ecological accounts will necessarily entail economists devising new global institutions (separate from the World Trade Organization, International Monetary Fund, and World Bank) that will create and then monitor sustainable finance, trade, and growth.

• Career Advice from the Expert

Using economic policy to change the world is still a fairly radical idea. It involves moving away from issue-specific approaches that may focus on dolphins, endocrine disruptors, or forests, and toward a solution-oriented approach that strives for accountability, changed prices, and different ownership structures. Taking apart the current money-driven world and turning it to achieve different ends is a difficult task. Prospective professionals should consider if they are ready to measure change in dollars and cents rather than directly in pounds of pollution avoided or acres preserved. Additionally, they should consider whether they want to be a bridge between policy wonks and grassroots communities. Translating economic and policy work for the public can be difficult, but the ultimate reward is a big idea backed by real people in real communities.

Prospective professionals should test their readiness by learning about and joining key debates in the field, and focusing on the economic dimensions of the environmental struggles that interest them. For example, an aspiring economist might figure out the real cost of driving in her home state of Mississippi by calculating the social, health, and environmental externalities. Aspiring professionals should also understand and stay open to economic practices that currently degrade the environment; today's harmful tools may hold the seeds of tomorrow's revolutionary approaches. For example, what can we learn from corporate bond traders about how to value forests?

More than anything, people interested in this work should constantly strive to make it real. The economy we've inherited, as well as current system of prices and economic rewards, manifests all the wrong things every day. So to work in this field, you must bring new and different ideas to the ground, where real economic impacts engage with real people and communities. Ideas and work cannot stand only on merit; they must belong to actual constituencies.

Accomplishing this goal is one of the most satisfying and powerful things you can imagine.

RESOURCES

Asian Pacific Environmental Network, www.apen4ej.org

Common Assets Alliance, www.commonassets.net

Environmental Justice and Climate Change, www.ejcc.org

Gund Institute for Ecological Economics, www.uvm.edu/giee

Indigenous Environmental Network, www.ienearth.org

International Society of Ecological Economics, www.ecoeco.org

Natural Capitalism, www.natcap.org

Polaris Institute, www.polarisinstitute.org

Redefining Progress, www.rprogress.org

Sky Trust Proposal, http://www.cfed.org/sustainable_economies/common_
assets/index_skytrust.html

8. ECOTOURISM

Martha Honey at Igazu Falls, Brazil.

A CONVERSATION WITH
Martha Honey

As ecotourism gained popularity worldwide, the United Nations named 2002 the International Year of Ecotourism. However, experts like Martha Honey say there is still much work to be done. "Ecotourism is an idea whose time has come, but it's too early to declare victory. For industry, there are still enormous amounts of work necessary to establish clear principles and practices for sustainability. For the traveling public, the challenge is finding environmentally and socially responsible companies that can provide a great holiday."

Honey is executive director of both The International Ecotourism Society (TIES) and the Center for Ecotourism and Sustainable Development (CESD). Having worked for twenty years as a journalist in East Africa and Central America, she has written hundreds of articles, reports, and books, including *Ecotourism and Certification: Setting Standards in Practice* (Island Press, 2002) and *Ecotourism and Sustainable Development: Who Owns Paradise?* (Island Press, 1999).

Honey obtained a Ph.D. in African history from the University of Dar es Salaam in Tanzania. Through her current work at TIES and CESD, she continues to demonstrate the need to use ecotourism as a development and conservation tool and to define international ecotourism certification standards.

What Is the Issue?

Tourism, widely touted as the world's largest industry, is designed for enjoyment and relaxation. For this reason, it has not been closely scrutinized like many other industries, including arms, oil, or drugs. But does tourism belong in the same category? Ecotourism experts would say absolutely. While tourism sells pleasure, in practice it has caused vast environmental, cultural, social, and economic damage to many areas of the world.

• Environmental Degradation and Local Exploitation

In the 1970s, we began to recognize that mass tourism and other traditional economic activities were causing environmental degradation on a global scale. For instance, extractive industries that exported goods to developed nations were degrading the rainforest in Latin America. We realized that areas like the rainforest needed preservation for the survival of both local indigenous groups and the entire planet. In East and southern Africa, wildlife was being destroyed even in national parks such as Serengeti and Maasai Mara due to an incredible level of poaching. As sophisticated international poaching operations incited warfare-like situations between poachers and park guards, animals such as rhinos and elephants verged on extinction, while indigenous communities saw no benefits from tourism. Rather, they often lived in squalor after being expelled from their homelands when the colonial and postcolonial governments created national parks. Poor people living on the parks' perimeters began to join poaching networks to make money in order to survive. Additionally, local communities were often forced to deplete the natural resources on which their livelihoods depended.

• Inequity and Leakage

Many of the most pristine and beautiful places in the world are also inhabited by some of the most marginalized people in the world, who usually don't have the skills necessary to become involved in international tourism. Historically, and particularly in developing countries, international tourism has been a business owned by wealthy people, often foreigners. This usually means that most profits leave the place where tourism occurs. The World Bank has estimated that in many countries, "leakage" of tourist dollars may be 50 percent or more. On some Caribbean islands as much as 85 to 90 percent of tourism profits either never enter or don't stay on the island; typically, the only money that does remain is used to pay low-wage workers. For this reason, this type of tourism has been a poor development tool. As such, leading development and lending agencies such as the World Bank and Inter-American Development Bank, which heavily pushed tourism as a development tool in the mid-twentieth century, closed their tourism offices in the late 1970s and halted conventional tourism lending.

How Are Environmental Professionals Approaching the Issue?

In the late 1970s, some people began experimenting with alternatives to mainstream tourism by asking how economic activity in the rainforest can be less destructive than the logging, oil drilling, or agriculture occurring there now. Further, environmentalists, **community organizers, park rangers**, and scientists began asking how these activities could be small-scale, locally owned, or run in partnership with nongovernmental organizations (NGOs) to provide alternatives so that tourism could be beneficial to the environment and local people.

Although alternatives to poaching and other destructive tourism impacts began emerging in the late 1970s, it was not until 1990 that The International Ecotourism Society (TIES) was founded and put forth the now widely accepted definition of ecotourism, which is "responsible travel to natural areas that conserves the environment and improves the welfare of local people." Additionally, ecotourism is travel that:

1. Is located in natural destinations
2. Minimizes impact
3. Respects local culture
4. Builds environmental awareness, both for travelers and people in the surrounding areas
5. Provides direct financial benefits for conservation in areas being visited
6. Empowers local people, enabling them to be directly involved in improving their quality of life
7. Supports human rights, democratic movements, and basic standards such as those set out by the International Labor Organization and various human rights organizations

Properly understood, ecotourism is not just a niche market of the tourism industry. It's a revolutionary concept calling for a new form of travel, where the seven defining principles are ultimately incorporated into all tourism functions. Ecotourism not only describes what the tourist does, but also the impact of travel—and it contends that this impact can be positive. Unlike conventional tourism, ecotourism should minimize environmental damage and keep much of the profits in local communities. In this way, ecotourism actually throws out a gauntlet and challenges conventional methods. Today, there are many different models for how ecotourism can be successfully used as a tool for environmental protection, community benefit, and empowerment.

- Forging Local-External Partnerships

Often, poor rural communities lack the knowledge necessary for successful ecotourism such as knowledge of another language, of the level of service that many tourists require, or of complexities of the tourism market. Additionally, local communities are not always interested in working in the tourism sector. Therefore, when community lands and resources are used for ecotourism purposes, economic principles deem it's necessary to find equitable compensation and to establish the rights and responsibilities of both the tourism business and local community.

In some places, NGOs and private ecotourism companies are working with local communities to help them acquire the skills to tap into and benefit

BOX 8-1

ECOTOURISM TOOLS FOR SUSTAINABLE DEVELOPMENT

- Forging Local-External Partnerships
- Developing Government Partnerships
- Changing Parks Management
- Providing Direct Conservation
- Establishing Eco-Certification and Eco-Labeling Standards
- Conducting Impact Studies

from ecotourism. This presents opportunities for people interested in sustainable tourism to work directly in ecolodges or with NGOs promoting ecotourism. In Costa Rica, for example, a popular choice for aspiring American ecotourism professionals is to work as interpretive or **naturalist guides** in public and private parks, or as ecolodge managers. While these opportunities can be exciting, the first priority should be to hire local residents, since job creation and strengthening local economies is an essential part of any ecotourism operation.

Ecotourism professionals also help locals assume positions as hotel employees and guides by providing language and interpretation training. NGOs, public-private partnerships, and governments can help provide this training, which is necessary for locals to enter the field at higher levels than simply cleaning rooms or waiting tables. Overall, these partnerships will help deliver financial benefits because money earned by locals will stay more readily in the community.

In South Africa, for example, white South Africans own many of the private companies involved in "boutique nature tourism," such as tented camps or luxury ecolodges offering small-scale, high-quality nature tourism. Since the end of apartheid, some of these companies have evolved into responsible ecotourism operations that provide economic and conservation benefits for local communities, and also hire and train locals as partners or co-owners. Wilderness Safaris is one company that co-owns several lodges with surrounding communities. These local communities are entitled to a certain percentage of lodge profits in return for use of the land, water, and other natural resources on which the overall tourism, lodges, and tented camps are built. Locals also supply labor and staff for running the facilities, while Wilderness Safaris puts in capital and marketing for the business. Together, this private company and the local communities with which it interacts have built several small ecotourism projects in South Africa.

- Developing Government Partnerships

Similar to private businesses, NGOs and governments can also partner with local communities to provide these types of benefits. An NGO in South Africa called Mafisa has worked with the local Makuleke people in Kruger National Park and the South African government to help the Makuleke get back land

NATURALIST GUIDE

AT A GLANCE

Employment:

20,000

Demand:

Rising

Breakdown:

Public sector, 75 percent

Private/nonprofit sector, 25 percent

Trends:

• More naturalists are interpreting history and culture as well as the natural world as the field grows more competitive.

• Ecotourism is growing rapidly, providing more positions all over the world.

Salary:

Starting salaries begin at $22,000, with median salaries ranging from $30,000 to $40,000 annually.

JOB DESCRIPTION

Naturalist or interpretive guides share their passion for nature by educating others about its wonders at nature centers and ecotourism destinations all over the world. Tim Merriman, executive director of the National Association of Interpretation in Colorado, likens these guides to shamans in native cultures who often knew much about the spiritual world to share with their people. "A shaman shares the stuff that was vital and necessary," he says, "and that's what we do. Interpreters transmit our understanding of nature and science to the next generation."

In the ecotourism field, the remoteness of this work might be considered a deterrent, although most guides thrive on spending their days outdoors introducing tourists to plant and animal life and a specific region's culture and history.

To cater to a variety of ecotourist needs, guides must use relaxed and informal approaches when discussing local ecosystems and slight habitat changes that can have vast impacts. For these reasons, it is important for naturalists to "subtly educate guests while integrating them with the natural surroundings in order to cause as little impact as possible," says Mike Hartman, owner of Tiamo Resorts in the Bahamas.

Naturalist guides also coordinate and organize trip itineraries and prepare financial reports. Their days begin early and can extend into evening hours as they work on administrative tasks or lead nighttime tours. Guiding can be seasonal work, although some operators work year-round depending upon location and climate. While they receive regular salaries, a large portion of guiding incomes depends on tips from ecotourists.

GETTING STARTED

- A bachelor's degree in conservation, fish and wildlife studies, forestry, wildlife biology, or field biology is recommended.
- Look for professional certification programs, such as those offered by the National Association of Interpreters at www.interpnet.org.
- Strong public speaking skills are important for presenting cultural and natural information in a dynamic manner.
- Knowledge of foreign languages, such as Spanish or French, is invaluable.
- Working or volunteering in local areas with organizations like the Student Conservation Association (www.sca.org) or asking to shadow a paid naturalist guide for a day will be useful in gaining experience in this field.

ADVICE FROM THE PROS

A naturalist for thirty years, Alan Kaplan supervises groups of youths for weekend camping activities and adult nature programs at the Tilden Nature Area in Berkeley, California. "What we try to do is educate people enough so they care about the natural world," said Kaplan. "The idea is, if you care about something, you'll care enough for it that you want to protect and preserve it." Kaplan advises finding a specific passion in the natural world, such as plant life or entomology, to acquire a distinct edge over other job applicants. "This way, you'll be the go-to person for that topic," he says. "You'll offer some little spark that shows that you can learn and absorb an entire subject matter."

that was seized from them by the white-run apartheid government in the 1960s in order to expand the park. The result of Mafisa and other organizations' involvement was that the government gave back land under the condition that locals would keep it for wildlife purposes, using it only for ecotourism.

This should be a win-win situation for the government and the Makuleke: the locals regain ownership of their land and the government ensures that the land continues to be protected. The Makuleke also benefit from stewardship of that land, because ecotourism is probably the land's most profitable use. All together, the South African government, Mafisa, private companies, and the Makuleke have worked to promote tourism that does not degrade the land or community. Mafisa has also continued its commitment to help empower the Makuleke and other poor communities by teaching them capacity-building skills necessary for understanding how to profit from tourism.

• Changing Parks Management

The Makuleke example also shows how ecotourism can help change the way parks are managed. In much of eastern and southern Africa, **park rangers**, park guards, scientists, and **community organizers** have begun to rethink parks management, arguing that wildlife will survive only if there is harmony, not hostility, between local peoples and parks.

Some national parks, for instance, now give a certain percentage of visitor gate fees to local communities for development projects such as school classrooms, grinding mills, and water pumps. Some tour operators and lodges are consciously hiring locals and giving them skills so they can be involved in tourism in ways that respect their culture. Local people are also promoting their culture and marketing crafts to foreign tourists. While not perfect, these policies have helped turn the tide. Poaching has gone down where local people are receiving significant benefits from ecotourism. The picture is not perfect, however: in far too many places, corrupt governments and irresponsible tourism operators continue to exploit local peoples and damage the environment while paying lip service to ecotourism.

- Providing Direct Conservation

One way that local people are benefiting from genuine ecotourism is that a growing number of tourism companies are providing financial and material support to host communities where they operate. In this trend, called the "travelers' philanthropy" movement, some ecotourism businesses are donating a share of their profits and soliciting contributions from guests to support various conservation and development projects in host communities.

In addition, a number of international environmental groups have ecotourism programs that promote conservation, such as the Rainforest Alliance, Conservation International, and the RARE Center for Tropical Conservation. Sustainable tourism or ecotourism departments at these organizations run training programs, conduct studies, purchase land, support parks, and in some cases run their own conservation and ecotourism projects in Asia, Latin American, and Africa. They employ environmentalists, **conservation biologists**, natural resource experts, and social scientists to implement conservation projects. Similar positions are also available for people who want to work for smaller or more local NGOs.

Conservation Corporation Africa (CC Africa) is one private South African company that has projects that provide benefits to host communities. To construct Phinda, a cluster of high-end luxury lodges, CC Africa bought old, degraded farmland from farmers in the Kwazulu Natal Province in South Africa and restocked it with wildlife, creating an absolutely magnificent, privately owned game park. In this case, CC Africa's "ecological engineering" of the land surrounding Phinda has benefited both the environment and local communities on the private reserve's periphery. Phinda also supports a variety of community projects, including schools and a small hospital.

Phinda is a classic example of restoration through reintroducing wildlife into an area that had been crippled by chronic drought and denuded by cattle grazing. This has been the case in much of southern Africa, where unproductive land has left locals with few options and only traditional wildlife like elephant, rhino, lion, and wildebeest can survive. Restocking and creating a game reserve for ecotourism is one of the few possible sustainable economic activities.

PARK RANGER

AT A GLANCE

Employment:

15,000

Demand:

Steady

Breakdown:

Public sector, 100 percent

Trends:

• Job security is often stable, as the economy's progress or decline seldom affects park rangers.

• Demand for rangers who collect fees and manage resources is greater than for those who work in educational and protection areas.

• More jobs will open in the next few years as park rangers start to retire or move into other fields.

Salary:

Entry-level salaries begin around $27,000, with annual median earnings reaching $45,000. Most employees earn between $35,000 and $55,000.

JOB DESCRIPTION

Although many nations around the world have developed park systems, they are particularly well-developed in the United States. The United States is home to nearly 400 national parks, 5,500 state parks, and thousands of local parks that offer many natural wonders and history. National parks such as Yosemite and Acadia introduce visitors to vast tracts of undisturbed wilderness. Other national landmarks, such as the Aztec Ruins National Monument and Gettysburg National Military Park, commemorate native cultures and Civil War battles, respectively.

Park rangers manage the conservation and use of natural, historical, and cultural resources. Park rangers "perform two major types of duties—resource protection, which involves law enforcement; and interpretation, which involves public education," states Bill Sanders, a board member of the Association of National Park Rangers (www.anpr.org).

Indeed, park rangers perform all kinds of duties, including natural resource management, law enforcement, forest and structural fire control, search and rescue operations, emergency medical services, interpretation and public speaking, conservation management, predator control, historic research, public relations, and even demonstrations of folk art and local crafts.

Dennis Knuckle, a park ranger at the gates of the Arctic National Park and Preserve in Alaska, says that his work varies with the seasons. During the winter, he takes care of administrative tasks, such as preparing the park's budget, scheduling maintenance work, and attending training sessions. In the summer, he spends most of his time doing field patrols, introducing visitors to the park's natural wonders and, at times, conducting search and rescue missions.

Park rangers spend most of their time outside, which means they must remain in good physical condition and be comfortable in all kinds of weather and working environments. In the United States, they work exclusively for the National Park Service or state parks. Elsewhere in the world, they may work for national or provincial governments, or even in private reserves.

GETTING STARTED

• A bachelor's degree in natural resources management, park and recreation management, history, archaeology, or natural sciences is required. Coursework in law enforcement is also valuable.

• Look for seasonal opportunities at state or local parks, or through the National Park Service (www.nps.gov).

• Most parks require certification in advanced first aid, CPR, fire control, and search and rescue procedures.

ADVICE FROM THE PROS

Bob Fuhrmann, a park ranger and educational program coordinator at Yellowstone National Park, says his job involves teaching the park's younger visitors about Yellowstone's wildlife, geology, and human history. "It involves opening people's eyes to what's going on around them," he says, referring to the park's hot springs, geysers, and other natural wonders.

"To work in this field," Fuhrmann continues, "you must enjoy learning about the outdoors, have a passion for nature and history, and have the desire and drive to teach people what they might not know about their country's wilderness." Fuhrmann also suggests that any volunteer or work experience in parks is a stepping-stone in this field.

· Establishing Eco-Certification and Eco-Labeling Standards

Because the tourism industry lacks basic standards for measuring sustainable operations, many companies will use the term "ecotourism" for marketing purposes, even though they may not engage in practices that reflect its true meaning and principles. They may not, for instance, make any tangible contributions to local community projects, or may, in drought-prone regions, siphon off large amounts of water to use on hotel lawns and laundries. In order to truly ensure that public or private tour operators, NGOs, governments, and community members are working to provide environmental protection and responsible ecotourism, many experts now advocate for certification systems and programs to set standards and measure the environmental, social, and economic impacts of tourism businesses.

There are a growing number of sustainable tourism certification programs designed to measure environmental, socio-cultural, and economic equity issues both internally (within business, service, or product) and externally (on the surrounding community and physical environment). These certification programs set criteria for measuring the performance of lodges and other types of tourism businesses. Through on-site audits, environmental certification specialists determine if businesses that voluntarily enroll in such programs comply with the criteria. Those that do are given a logo or green label to use for marketing purposes.

Two of the best-known certification programs are Costa Rica's Certification for Sustainable Tourism (CST) program and Australia's Eco Certification Program (formerly known as the Nature and Ecotourism Accreditation Program [NEAP]). While CST certifies Costa Rican hotels and lodges, the Eco Certification Program assesses Australian ecotourism tours, attractions, and accommodations, mainly in the Queensland area. Both CST and the Eco Certification Program are being used as models to create other nationally based certification programs in Asia and the Americas.

In recent years, we've also seen many codes of conduct for how tourists should behave if they go into indigenous areas or interact with indigenous communities. For example, Matemwe Bungalows, a small lodge on the Tanzanian island of Zanzabar, provides a little sheet in each room that describes the local Muslim community and outlines some "do's and don't's" for tourist behavior and dress. NGOs, guidebooks, ecotourism societies, governments, and private businesses have also put out codes of conduct for how travelers should behave, particularly for people interacting with indigenous communities. In addition, journalists can play a vital role in promoting sound ecotourism and exposing frauds through solid investigative reporting. Monitoring and reporting help determine what is actually occurring, both in terms of uncovering "greenwashing" scams masquerading as ecotourism and also in terms of highlighting exemplary projects.

- Conducting Impact Studies

Although ecotourism standards are still being developed, NGOs, governments, private businesses, and international agencies must often perform impact studies to assess an operation's environmental impacts. For example, international agencies like the World Bank, Inter-American Development Bank, and USAID typically require environmental impact studies before any project is built. Many ecotourism experts now advocate that these studies be expanded to also measure social and community impacts so they become part of the legal requirements for ecotourism projects. Researchers can design and carry out such impact studies to determine how the practices of ecotourism lodges, operators, and other businesses affect local communities and ecosystems. This research can help determine if ecotourism is benefiting communities and conservation, as well as the business itself.

Career Advice from the Expert

Students interested in pursuing an ecotourism career should obtain foreign language, business, marketing, and fundraising skills, and take courses in sustainable tourism. Increasingly, there are a number of universities and colleges around the country that offer courses or degrees in ecotourism or related fields, including George Washington University, Stanford University, Texas A&M, West Virginia University, Humboldt State University, Johnson State College, Paul Smith College, and the University of Florida.

Aspiring ecotourism professionals should have at least a bachelor's degree and preferably a master's degree in tourism, hotel management, international development, or environmental-related disciplines such as biology, conservation, or ecology. Additionally, backgrounds in social science disciplines

Career Spotlight

Wildland Adventures, Inc.
Seattle, WA

Kurt Kutay loves his job. As founder and president of Wildland Adventures, he brings the world of nature to thoughtful tourists and contributes to the creation of a sustainable economy. In addition to writing articles for popular magazines and guidebooks on adventure travel and ecotourism, he also travels to some of the world's most beautiful wild places every year—and makes a pretty good living in the bargain. Nice work if you can get it!

Educated with a B.A. in economics from the University of Oregon in 1975 and a master's degree in natural resource management from the University of Michigan School of Natural Resources in 1982, Kurt subsequently worked for the Costa Rican National Park Service. "When I returned from Costa Rica in 1983, I saw the opportunity to tap into a growing interest among travelers who wanted to learn about rainforests and see wildlife in tropical environments." He started taking people to Costa Rica with the intention of using his travel business to support conservation and help rural people living around protected nature reserves. Wildland Adventures was born in 1986.

Located in Seattle, Washington, the company has established itself as a recognized leader in ecotourism. It offers active, outdoor nature and culture explorations to destinations throughout the world. As an "outbound tour operator," its core staff of eight people is primarily involved in designing and marketing adventure tours, contracting with local "inbound ground operators" to perform tour services, and working closely with its travelers to choose vacations and prepare for the adventure of a lifetime.

"Although I never studied business or knew much about the travel industry, I learned along the way and thoroughly enjoy the craziness of being a business owner," Kurt relates. Daily tasks involve oversight and involvement in financial management, sales and marketing, business development, personnel management, graphic design, and information technology.

The principal position of other staff in the company is "destination program director." These professionals are responsible for sales and administration of group and individual travel arrangements to South America, Central America, the Middle East, Africa, Australia, and other parts of the world. Unlike the tasks of travel agents or more conventional tour companies that sell packaged tours, sending travelers to remote undeveloped areas requires detailed knowledge of the destination, ecosystems, indigenous cultures, and local conditions.

Nicole Harrison is the woman to talk with if your dream vacation land is Africa. Wildland's "Africa Program Director" was hired sight-unseen when she called the company in 1998 as she was finishing her studies in environmental studies at George Washington University. She had worked as a counselor in an outdoor environmental education center and had also spent six months at the Center for Field Studies as an undergraduate assisting with research on elephants in Kenya. Although aspects of in-country tours are often in the hands of local contractors, Nicole has been on safaris all over the continent. She has led groups of eco-tourists as they climbed Kilimanjaro, encountered gorillas in Uganda, canoed down the Zambezi, and tracked game in the Luangwa Valley of Zambia.

Daily responsibilities of program directors include customer sales, client predeparture services, local ground operator communications, client account management, preparation of sales materials, and designing new itineraries and programs. Experience in tour operations, administration, and adventure travel field experience is important for this work. It also helps to be detail-oriented, well-organized, and incredibly versatile, and to have an enthusiastic personality and professional telephone presence.

Some staff are more office-bound. Shamus Daily, Wildland's information technology specialist, made himself irre-placeable when he was hired full-time in 2002 after working part-time for the company while completing study at a local community college in web development, networking, and digital information management. "Like most businesses, much of what we do boils down to information management," says Shamus. "Using technology creatively is essential for reaching customers in a growing industry."

That growth is a source of some concern for Kutay and his staff—not for the company, but for the wild nature they love. "Ecotourism is ultimately about conservation and helping local people make a living. It's so important to do this work correctly, so that we don't damage the very ecosystems and communities we're trying to save."

Wildland Adventure guests get up close and personal with elephants on an African safari.

such as anthropology or architecture and studies that focus on specific areas such as Africa or Latin America are useful, depending on a person's desired specialization.

Most importantly, prospective professionals should seek long-term overseas work experience. Whether this work is volunteer or paid, it should relate to developing ecotourism or community-based tourism initiatives; integrating conservation, development, or income-generating projects including handicrafts; or guiding tours or expeditions. As the ecotourism field is increasingly moving toward community-based and eco-certification initiatives, any direct experience in community development or certification—even if it does not involve tourism per se—will be beneficial. Experience in participatory planning and appraising ecotourism operations and destinations will also help prepare for work in this field.

Organizations that offer internships related to these types of ecotourism projects include The International Ecotourism Society (TIES), the Center on Ecotourism and Sustainable Development, Conservation International, The Nature Conservancy, Rainforest Alliance, and World Wildlife Fund, as well as several other major environmental and development organizations.

Overall, poor communities in Africa, Latin America, and Asia (as well as development and funding agencies such as the Inter-American Development Bank and the World Bank) are increasingly using ecotourism as an important tool for conservation and poverty alleviation. In reality many ecotourism projects around the world fail due to the lack of proper management, marketing, and service. There is, therefore, a need for skilled professionals willing to work with local communities, development agencies, or NGOs to help build projects that conform to ecotourism's basic principles. Accordingly, a good way to start your career is to spend a summer or semester volunteering on a rural community project or with an NGO like one of those already mentioned.

RESOURCES

Center on Ecotourism and Sustainable Development, www.ecotourismcesd.org
Certification for Sustainable Tourism, www.turismo-sostenible.co.cr
Conservation Corporation of Africa, www.ccafrica.com

Green Globe, www.greenglobe21.com
Nature and Ecotourism Accreditation Program, www.ecotourism.org.au/neap.cfm
Rainforest Alliance, www.rainforest-alliance.org
RARE Center for Tropical Conservation, www.rarecenter.org
The International Ecotourism Society, www.ecotourism.org
Wilderness Safari, www.places.co.za/html/wilderness_safaris.html
Wildland Adventures, http://www.wildland.com/
World Tourism Organization, www.world-tourism.org

9. ENERGY

Worldwatch senior fellow Seth Dunn.

A CONVERSATION WITH
Seth Dunn

Because it relies on exhaustible and polluting fossil-fuel sources, it's clear that our current energy system needs an overhaul. "Transforming our energy system is the single most important change we can make if we're serious about protecting the environment," says Seth Dunn. "For the first time in a while, we are able to identify realistic actions for accelerating that transformation, making now a very exciting and significant time."

Dunn is a senior fellow at the Worldwatch Institute. He has proposed "deliberate decarbonization," or the progressive reduction of the amount of carbon produced for a given amount of energy, as the first and foremost step in shifting the current energy system away from dependency on fossil fuels. Reducing this dependency will not only produce positive environmental benefits, but will also create millions of jobs, lessen geopolitical tensions that can lead to war, and prepare us for the energy age beyond oil, coal, and gas—which is sure to come, whether we're ready or not. A highly sought expert, Dunn has participated in four rounds of United Nations Climate Change negotiations and expert meetings of the Intergovernmental Panel on climate change. He obtained his M.B.A. and master's in environmental management from Yale University and has published more than sixty reports, book chapters, articles, and op-eds, including "Decarbonizing the Energy Economy" in *State of the World 2001* (Norton, 2001).

What Is the Issue?

We've known for a long time that our energy use has environmental impacts. But only lately have we come to understand the scale and magnitude of the problems our energy dependence is creating. In the twentieth century, our economy's ever-growing hunger for fossil fuel has moved the ecological consequences of energy use from local to regional and global levels, to the point where it is altering the earth's atmosphere. Compounding that fact is the energy system's considerable inertia. Changing the way we use energy is like turning around a supertanker to avoid an iceberg. We can see the iceberg—but it's not quite clear how far away it is, or whether we can avoid it.

• Environmental Consequences of Energy Use

The local effects of energy use relate primarily to particulate emissions and their health impacts. Concerns about the implications of coal burning date back to the seventeenth century, and they came to a head in the twentieth century with the infamous coal smogs of London and Pittsburgh. Such effects were often viewed as the price of progress; in a modernizing society, smokestacks and smog were seen as the accepted trade-off for economic development. Today, however, there is growing awareness that this model may not be sustainable for China, India, and cities in other countries in the developing world.

The traditional tack for dealing with energy's environmental side effects has been an "emergency room" approach: pollutants are dealt with individually, and only after they become a problem. Power plants in the United States built higher smokestacks to deal with local particulate pollution. The unintended consequences were that pollutants were carried aloft in the atmosphere and then fell as acid rain in sensitive regions such as the Adirondacks. Today, much of industry is excited about "carbon sequestration," where carbon dioxide emissions will be stored underground. If history is any guide, however, this end-of-pipe approach will prove more costly and less environmentally effective than anticipated.

• Reliance on Fossil Fuels

The fundamental problem underlying the current energy system is its overwhelming reliance on fossil fuels as a primary source. The bulk of global commercial energy consumption is derived from the combustion of coal, oil, and natural gas. These sources may be sustainable for a world population of one billion, but they will not be sustainable for a population of eight to ten billion. While end-of-pipe pollution control has achieved some progress with respect to addressing local air pollution and acid rain, the problem of global climate change necessitates a shift toward low- and no-carbon sources. We are not running out of fossil fuels per se. Instead, we are running out of the earth's absorptive capacity to assimilate carbon dioxide emissions. Even with a rapid transition away from fossil fuels, the emissions of the last 150 years have already committed us to several degrees of warming over the coming century, regardless of what we do.

• Energy Poverty

Another unsustainable aspect of the current energy system is the widespread disparity in access to modern energy services. Right now, roughly two billion people worldwide lack access to any modern energy services, most of them in rural areas that make them unlikely to gain energy by grid extension. Most of these people are also poverty-stricken and unable to gain necessities for basic electrical needs. Alternative power sources are much needed in these areas so people can prevent further fossil-fuel-based generation and preserve their environments.

How Are Environmental Professionals Approaching the Issue?

The main culprit when it comes to the environmental impacts of energy use is carbon. But the good news is that, throughout history, the global energy system has been gradually decarbonizing, or reducing its dependence on carbon as an energy resource, as humanity has moved from wood to coal, and then to oil and natural gas. The next logical step toward carbon-free energy is to move toward renewable energy and hydrogen, but this will require several decades of sustained commitment.

• Decarbonizing Our Economy

Decarbonizing the energy system requires creating processes that incorporate the "three Ds": decentralization, downscaling, and diversification. The new system will be decentralized in terms of decision making because consumers will have increasingly more control over their power sources, making them independent from utilities whether or not they go off the power grid. This will create democratization, where people can help shift the system by showing demand for alternative energy such as wind power or solar. However, we need to create more environmental awareness in order to make changes.

Technology will become decentralized through increases in downscaled, smaller "micro power" systems like fuel cells, micro turbines, small-scale natural gas systems, and renewable energy sources. These small-scale technologies will lead us away from multimegawatt systems and toward smaller units that are more cost-efficient, perhaps even available to put in basements or on rooftops. Finally, the new system will be diversified, simultaneously involving solar, wind, and other sources, creating a real mix depending on geography and how far each technology has developed. These three components will distinguish the twenty-first-century energy system from its twentieth-century predecessor.

The imperative of creating a new energy system presents exciting opportunities for environmental professionals, whether through work in the alternative energy industry or by effecting change through nongovernmental organizations (NGOs) or government. This is an immature area, but its long-term prospects are promising, as professionals are working to approach the problem through technology and business, policy and politics, and public education.

- Building a New Energy Business

The "alternative energy business" in 1900 was a new fuel called petroleum that was being adopted in niche markets as an initial replacement for kerosene and lanterns. Today, of course, the petroleum business is among the world's

BOX 9-1

TRANSFORMING THE ENERGY SYSTEM

- Decarbonizing Our Economy
- Building a New Energy Business
- Using Technology as a Tool for Sustainable Development
- Rethinking the Business Side
- Developing New Policies and Politics
- Educating the Public

largest industries. Back then, the rate of oil use was increasing at roughly the same rate now being experienced with a number of today's renewable energy resources, and with similar economic impacts. As then, there are many career opportunities opening up through hundreds of eager business start-ups, continuous mergers and acquisitions, and investment successes and failures.

Wind Power

With an average annual growth rate of about 25 percent over the past ten years, wind power has now become a lively market of approximately $3 billion due to rapid cost reductions and policy incentives. Most growth is occurring in Europe thanks to very supportive government policies—"feed laws" that guarantee developers fixed prices for selling wind power to the grid. But developing countries such as India and Brazil are also experiencing wind power growth. Wind power uses mostly decentralized systems, clustered in large farms or scattered in smaller groups. These power systems are not the windmills of the nineteenth century in the U.S. Midwest. Rather, they are advanced aerodynamic fiberglass plates that maximize and convert wind speed to power generation. This energy source has minimal environmental impact and no direct emissions. It is also generating a wide variety of entirely new careers. In addition to technical positions for electrical engineers and wind power technicians (for installation, maintenance, and repair), the expansion of the wind industry requires energy and **policy analysts** to develop and tweak emerging incentive policies. Marketing, sales, finance, and business management professionals are also needed to help growing firms prosper in an exceptionally demanding marketplace.

Solar Power

The solar photovoltaic (PV) market is spreading mostly overseas, especially in Asia. While solar power products are expensive, their efficiency is improving, and we are able to convert high percentages of sunlight into electricity, meaning the products ultimately will require fewer materials. Solar power illustrates how intelligent design can make alternative power sources more attractive. For example, some solar products are being integrated into buildings, rooftops, and window materials so they can offset building costs and make buildings and building materials more energy independent. Specifically, green building design is increasingly using solar PVs as resources in the design process. One side of the One Times Square building in New York City, for example, is made of solar heated glass coated with translucent solar PV windows.

Like wind energy, solar power is also generating many new jobs at utility companies, solar product companies, research labs, and advocacy groups nationwide. A quick glance at the employment opportunities listed at Solaraccess.com, an Internet business and database that provides all information related to the overall renewable energy field, shows demand for PV systems designers in Vermont, solar products engineers in Los Angeles, technical sales people in Colorado, renewable energy program managers in New Mexico, and solar technicians in New Hampshire.

Geothermal Energy

Geothermal technology uses heat under the earth's surface to draw steam to use as an alternative energy source. Most geothermal systems use this steam to compress and dry gases, which is necessary to create electricity to drive a turbine. Geothermal heat pumps can be used on smaller scales to provide energy for uses like space heating in houses and buildings. This type of energy can also be used directly for heating. In the United States, geothermal initiatives have occurred mainly on the West Coast. Calpine, for example, has utilized the geysers in California as a productive geothermal energy source.

Biomass Energy

Biomass is another underutilized energy source. Agricultural wastes ranging from sugarcane bagasse to rice hulls can be gasified and turned into combustible fuels. Ethanol from sugarcane, for example, already supplies half of Brazil's automotive fuel. Additional biomass derivation methods may be realized from ongoing research and development.

Fuel Cell Technology

Hydrogen is the lightest and most abundant element in the universe. Using it as an alternative energy source will be vastly less damaging to the environment than our current use of fossil fuels. Most discussions about using hydrogen refer to fuel cell technology. Figure 9-1 explains how fuel cells work by using electrochemical devices that split water into hydrogen and oxygen. As the diagram shows, the energy process relies on electrochemistry and produces no direct carbon pollutants.

The challenge of utilizing hydrogen in fuel cell technology is reducing its cost. However, when companies determine how to mass-produce any product, a "model T effect" often occurs, where large costs decline as the volume of production

ALTERNATIVE ENERGY SPECIALIST

AT A GLANCE

Employment:

11,000

Demand:

Rising

Breakdown:

Private sector, 70 percent

Public sector, 30 percent

Trends:

• As wind and solar energy technology are perfected, these fields will become more privatized, creating more job opportunities with private companies.

• Most professionals start as entry-level technicians—jobs that are readily available—before earning more complicated engineering positions.

Salary:

Entry-level salaries begin around $40,000, with median annual incomes reaching $48,000 and top salaries around $65,000 and above.

JOB DESCRIPTION

As natural resources such as petroleum deplete, industry is turning to alternatives including wind and solar power. In both of these industries, alternative energy specialists design and install systems that use renewable energy sources.

In the solar energy field, for example, product and systems engineers create and design materials such as glass, thermal, and photovoltaic panels that glean the sun's energy to generate electricity and heat.

In the wind energy field, meanwhile, one of the largest-growing career opportunities is field engineering. Field engineers mainly install and maintain wind turbines that supply electricity and pump water. "This is the fastest-growing energy resource," said Christine Real de Azua, spokeswoman for the American Wind Energy Association (www.awea.org). "It has the potential to become a major power source."

In their daily work, field engineers and the technicians working under them must choose efficient sites for wind farms, since turbines must be placed in areas with enough wind for power generation. Once they install wind turbines, engineers then work with utility companies to ensure that local residents receive the electricity generated from the turbines. "The hard work results in a good wind turbine that runs well and pumps out electricity," says Walter Hornaday, a field engineer working in Texas. As the wind energy field grows, more field engineers and technicians will be called upon to install and manage wind turbines in developing countries, where millions of people live without electricity.

GETTING STARTED

• A bachelor's degree is required for employment. Many alternative energy specialists have backgrounds in mechanical, aerospace, or electrical engineering.

• Since wind energy is a relatively new field, experience may be difficult to find. Working with an electrical or aerospace engineer provides valuable experience.

• Many companies offer training programs for students, such as General Electric's Student Training Employment Program (www.gepower.com/about/careers/en/co_op_intern/index.htm) and Sandia National Laboratories' student internships in New Mexico (www.sandia.gov).

ADVICE FROM THE PROS

James Johnson, an engineer at the National Wind Technology Center in Boulder, Colorado, supervises wind turbines on 305 acres, where he performs tests and research in wind technology. A licensed mechanical engineer, Johnson says he learned about wind technology by working with engineers in the field. "There's a great deal more to learn about wind power than just in college classes," he says. "People who are successful in this industry have a strong environmental background, as well as an engineering one."

Johnson also says newcomers to the field should expect to work in a lower-level capacity at first before moving on to engineering, which involves much research. "We do a lot of research, especially atmospheric forecasting and looking into the technical aspects of developing wind technology, so you've got to like to work with scientific data."

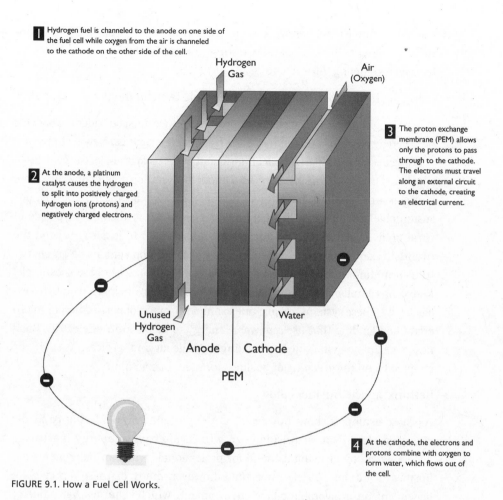

1 Hydrogen fuel is channeled to the anode on one side of the fuel cell while oxygen from the air is channeled to the cathode on the other side of the cell.

Hydrogen Gas

Air (Oxygen)

3 The proton exchange membrane (PEM) allows only the protons to pass through to the cathode. The electrons must travel along an external circuit to the cathode, creating an electrical current.

2 At the anode, a platinum catalyst causes the hydrogen to split into positively charged hydrogen ions (protons) and negatively charged electrons.

Unused Hydrogen Gas

Water

Anode   Cathode

PEM

4 At the cathode, the electrons and protons combine with oxygen to form water, which flows out of the cell.

FIGURE 9.1. How a Fuel Cell Works.

increases. Some companies such as General Motors are hinting that they might mass-produce hydrogen fuel cells by 2010, indicating that we can expect exponential growth in these technologies and hydrogen use over the next several decades.

A hydrogen economy that has virtually no pollutants will decrease our ecological footprint and increase our ability to preserve and restore our natural surroundings. Although this transition will take several decades, global and local societies and quality of life will improve. Iceland has recognized this, and is aiming to become the world's first hydrogen economy by 2030. Its government has teamed up with academic institutions and the multinational companies Shell Hydrogen, Daimler-Chrysler, and Norsk Hydro to facilitate its transition from a fossil-based economy to a non-fossil-based economy. The consortium plans to make Iceland a testing ground for hydrogen vehicles and hydrogen-refueling infrastructure.

- Using Technology as a Tool for Sustainable Development

When discussing energy efficiency improvements, we must consider the benefits of lowered emissions and reduced costs from energy savings. New technologies —even current ones like compact fluorescent bulbs or hybrid-electric vehicles— can "buy time" for the broader transition to a carbon-free economy. These changes are small but significant, and can also be used as a tool for sustainable development. Bringing new technologies to the poorest people in rural areas of developing countries will enable them to leapfrog beyond the twentieth-century system, which means the social implications of an energy transformation will be most profound in the developing world. For example, Kenya has implemented solar pump systems that are helping people access electricity. These systems are just one example of many opportunities for micro credit and lending that use renewable energy to help countries electrify. Businesses will play a vital role in providing these goods and services. As such, see chapter 15 on the greening of business for more information.

- Rethinking the Business Side

As these examples show, businesses are important drivers of change in the energy system. Not surprisingly, growth in alternative energy markets is accompanied by growing demand for professionals who can help implement business solutions. At the same time, however, many of these markets are dependent on a favorable policy environment, which requires that industry professionals also be aware of and involved in the policy process.

- Developing New Policies and Politics

The "three Ds" of transforming the energy system (decentralization, downscaling, and diversification) cannot be accomplished without political circumstances that support them. Currently, the bulk of public support favors the incumbent system, making it very difficult to bring about change; today's system provides $300 million in subsidies for fossil-fuel imports worldwide each year. We won't be able to transform this system until we start paying the real costs of energy by eliminating unjust subsidies and creating pricing structures that include environmental and social costs. These costs are currently not taken into account at gas pumps or in electricity bills. Additionally, U.S. energy legislation tends to be so insufficient and laggard that, by the time legislation is ready for a vote, laws have been dumbed down or changed so much that they do not provide the environmental and economic benefits they initially intended. The victims of this chronic impasse are the infant industries—solar and wind—that still rely on subsidies to maintain even a faint semblance of an equal footing in the marketplace.

So decision makers and **elected officials** should create policies to remove subsidies, tax breaks, and direct payments to producers of harmful fossil fuels. One new and promising market-based policy is the renewable portfolio standard (RPS), which requires utilities to devote a fixed and growing percentage of their electricity purchases to renewable energy. Many states, including Texas, California, Arizona, and Nevada, have passed a version of RPS. The Texas RPS is the most successful and has spurred significant growth in wind power development in the state. Taxing environmentally destructive systems in this way makes market energy prices more accurately reflect the full cost of energy, including the often degrading impacts of energy use on the environment.

However, any type of transition away from a fossil-fuel economy will not occur without the cooperation and involvement of all stakeholders who use and support the current energy system. Local, state, regional, and national governments, industries and businesses, and civil society must all be participatory stakeholders in a new system in order for it to succeed. We must develop more sophisticated market signals in order to reap the long-term economic benefits of an energy transformation.

Market signals are information that investors and businesspeople use to make decisions about when and where to spend money on new equipment, exploration, mergers and acquisitions, hiring, research, and development. Perhaps the most critical market signal is the differential between the cost of securing energy, the cost of preparing it for market by refining oil into gasoline, and the price at which it is sold. In many ways, prices are the information on which the entire private enterprise system makes decisions. As we've seen, however, prices usually don't reflect environmental, social, and economic security costs, making them crude market signals, especially in the energy sector. Government in particular can start by supporting and changing subsidy and tax policies. Also, if governments take less of an energy ownership role and instead take on a rule-setting role, they will become referees rather than

POLICY ANALYST

AT A GLANCE

Employment:

4,500

Demand:

Slow

Breakdown:

Public sector, 50 percent

Private and nonprofit sectors, 40 percent

Academia, 10 percent

Trends:

• Many nonprofit groups will require the expertise of policy analysts to advocate for their causes, creating demand for policy analysts in the next few years.

• Professionals with experience working on a political campaign or for an advocacy group are often attractive job candidates.

Salary:

Salaries range from $29,000 to $37,000 for entry-level positions. Median wages are $52,000, with top analysts averaging $80,000 and above.

JOB DESCRIPTION

Policy analysts seek opportunities to shape and create policies that preserve and protect the environment. They often possess a strong understanding of the relationship between the natural world and the social sciences, since their work affects both environmental and public health. "Analysts bring evidence and insight to bear on environmental policy for the public good," says John Hird, professor and director of the Center for Public Policy and Administration at the University of Massachusetts at Amherst. In academia, Hird focuses his work on environmental risk policy and justice, particularly regarding polluted sites near poverty-stricken and minority residences. "This is a large enough, wide-ranging field that you can make a difference in many areas of interest," he says.

A large part of a policy analyst's job is to keep up with current political, social, and environmental issues. To accomplish this goal, policy analysts must first understand the way political systems work. They also spend much of their time reading professional journals, newspapers, and popular literature. They research data, analyze policies and laws, and document their findings in coherent reports.

Analysts are always looking to steer policy on a sustainable path by working with many groups. Their work can be highly political. They constantly meet with other analysts, politicians, and citizens groups at conferences, government hearings, and community meetings to debate issues and look for solutions. Dr. Kenneth Green, director of the environmental program at the Reason Public Policy Institute, says his work is based upon effecting changes in public opinion and envisioning long-term change. To work toward these goals, he writes briefs, policy reports, and fact sheets to prepare for meeting and speaking with different groups.

Policy analysts are also realistic thinkers who determine whether proposals are feasible and doable within a reasonable amount of time. Besides academia, policy analysts work for consulting firms or as lawyers for nonprofit groups. Some also work for the government, directly influencing the administration.

GETTING STARTED

• A bachelor's degree is required in studies such as environmental or social sciences, with training in risk management and negotiation skills.
• A master's in public policy, public or business administration, or economics is highly recommended.
• Gain an understanding of political systems by working or volunteering for local and state government agencies, political campaigns, and environmental groups.

ADVICE FROM THE PROS

Kevin Shively, policy director for the Transportation Choices Coalition, says he advocates for legislation that will improve public transportation use in Seattle. "I think the most important thing in this field is listening to what your opponents are saying and seeing if you can find a middle ground," he says. "Keep abreast of what's going on everyday because you can't afford to fall a step behind."

Career Spotlight

Silicon Valley Power
City of Santa Clara
Santa Clara, California

The residents of Santa Clara, California, get 26 percent of their electricity from renewable energy, especially geothermal power. That's more than twice the California average of 12 percent from renewables and even higher than the national average that remains stuck in the single digits. When added to the power the city receives from large hydroelectric dams, nearly 70 percent of Santa Clara energy is provided by earth, sun and water—instead of coal, oil, and nuclear sources. Maintaining a steady flow of this environmentally sound energy at competitive rates is a daily task for the 138 city employees of Silicon Valley Power (SVP).

"As a not-for-profit municipal utility, we are community-focused," explains Larry Owens, division manager for customer service. "Our city leaders have decided to emphasize the benefits of renewable resources in the utility's power mix. We're committed to providing environmentally friendly power while maintaining low, stable rates. As a result, our customers enjoy unusually clean energy for less than eight cents per kilowatt-hour."

Joyce Kinnear, who has the intriguing title of "public benefits coordinator," plays an important role in ensuring that the city reduces its energy consumption and everyone can afford to turn on their lights. Her position was created in response to state of California requirements that all utilities collect and spend money for "public goods," even as the state loosened rules on the electric industry as part of overall deregulation policy. Today, Kinnear's portfolio includes over twenty programs, such as free energy audits; rebates for energy-efficient lighting, appliances, and equipment; incentives for efficient construction and remodeling; low-income assistance; solar energy demonstration projects; and an electric hybrid bus service.

Kinnear has a strong utility background, with previous jobs as manager of rate analysis for the city of Las Cruces, New Mexico, and senior rate analyst for PG&E Energy Services. She holds a master's degree in public administration from New Mexico State University. Ensuring public benefits to all city residents keeps her incredibly busy. "I'm responsible for program design, budgeting, staff resources, marketing, communications, special events, and lots of reporting," she says. "Getting everything done means hiring and managing project contractors in addition to my in-house staff."

That staff includes energy conservation specialist Leslie Brown,. Brown started her career at SVP as an intern while finishing her B.S. degree in environmental studies with an emphasis on energy at San Jose State University. Among other things, Brown staffs the Energy Conservation Hotline, answering questions for environmentally conscious residents who also want to save significant dollars on their electric bills. Brown's real passion, however, is solar energy. She manages all of the city's photovoltaic installations and directs "Neighborhood Solar," which collects monthly donations from customers to defray solar project costs. "The first system will be installed at an elementary school in 2004," Brown says proudly.

Generating electricity, even "clean" power, requires compliance with a dizzying array of environmental laws. William Reichmann, senior electrical generation engineer, learned his trade on the job through over twenty years of increasing duties. "We have city-owned power plants and third-party facilities that generate 57 megawatts (MW) from thermal power, 23 MW from hydro, and 20 MW from wind.

Damon Beck and Joyce Kinnear show SVP customers how to save money with energy-efficient appliances.

All have operating permits, and I'm responsible for meeting the requirements." Reichmann reels off an alphabet soup of federal (EPA, FERC, USFS), state (CEC, CARB), and local (BAAQMD, RWQD, planning, fire) agencies and laughs. "I try to bring some reality to the process."

In addition to more than 40,000 residential customers, SVP has 8,000 business and industrial clients. Keeping a select group of these firms happy is in the job description of key customer representative Damon Beck. Like Leslie Brown, Beck started as an intern while finishing a B.S. in environmental studies with an emphasis in energy management from San Jose State University. When California's deregulation compelled municipal utilities to develop strategies for holding onto critical business customers (and courting new ones), Beck's job was created. With deregulation somewhat on hold, he now focuses exclusively on meeting the needs of SVP's largest customers and working for the environment at the same time. "I advocate for energy-efficient conversions that keep costs down and improve our customer's competitiveness," Beck says. "When that happens, a difference has been made for the better."

players. This change will allow companies to compete in the marketplace without worrying about favorite policies that do not apply uniformly to all companies.

Educating the Public

In citizen preference polls for energy sources in the United States, the consistent top choices are renewable energy sources. This ranking is often an inverse proportion to the federal money allocated, signaling that government priorities need reorientation. This ranking may also be in opposition to people's awareness of their own energy usage. Look around your house—what choices do you have right now in terms of where to get power? You may be surprised to find that your energy comes from a nuclear power plant upstate, or perhaps from a coal or natural gas-fired power plant across town.

In the new energy system, you may be able to choose your energy source by deciding you'd like solar energy or that you want to buy from a green power company that puts half of its money into 100 percent renewable energy. The need for consumer influence is great, and it is important for **community organizers** and public advocates to build a public constituency of people who demand alternative energy, thereby providing momentum for implementation of technology and legislation. This will help the overall energy transition, although government regulation is ultimately required to ensure that people know their alternative energy products are certified green, similar to the way organic foods are labeled.

Career Advice from the Expert

Recent college graduates looking to help transform the energy system should become conversant in the industry by familiarizing themselves with appropriate publications such as *Wind Power Monthly, Solar Today, Renewable Energy World*, and *Hydrogen & Fuel Cell Letter*; websites and professors are also excellent sources of information. Understanding the current system's structure and how it is affected by policy is essential for becoming involved in opportunities to create change. If someone is interested in wind energy, for example, having familiarity with the industry's history since the 1980s is necessary. Additionally, prospective energy professionals should refine their focus and establish contact with relevant organizations in the private, nonprofit, or public sector, according to their interest. For example, in the case of an interest in wind energy, prospective professionals would be wise to become familiar with the work of the American Wind Energy Association.

Numerous NGOs, agencies, and trade associations in Washington, D.C., address energy-related issues, such as the American Council for An Energy-Efficient Economy and the Renewable Energy Policy Project. These and similar organizations can be an excellent starting point for developing an understanding of the complexity of energy policy and the varied interests of the many players who attempt to shape it. Many people leverage their policy knowledge into interesting positions within the private sector, at corporations such as BP Solar, GE Wind Energy, or Shell Renewables; conversely, some people with private sector experience segue to the public and nonprofit sectors. Prospective energy professionals should be persistent in their job searches, as an intrinsic interest in the field is important for maintaining contacts that may lead to securing a job.

While there is no particular academic requirement to enter the energy field, engineering and economics backgrounds are common. As new energy technologies reach or near commercialization, employers are increasingly showing demand for business skill sets including strategy, finance, marketing, and operations. Additionally, building cross-sector and cross-functional experience will give an aspiring alternative energy professional multiple perspectives. Being able to integrate the economic, financial, and policy dimensions of energy issues is an important asset. The resources section of this chapter provides a sampling of private, nonprofit, and public organizations an aspiring energy professional might use as a starting point for entering the field.

RESOURCES

American Wind Energy Association, www.awea.org
BP Solar, www.bpsolar.com
Ballard Power Systems, www.ballard.com
GE Wind Energy, www.gewind.com
Geothermal Energy Association, www.geo-energy.org
Icelandic New Energy Ltd, www.newenergy.is
National Hydrogen Association, www.hydrogenus.org
Renewable Energy Policy Project, www.repp.org
Silicon Valley Power, http://www.siliconvalleypower.com
Solar Energy Industries Association, www.seia.org
U.S. Department of Energy, www.energy.gov
Worldwatch Institute, www.worldwatch.org

10. ENVIRONMENTAL EDUCATION

Whitney Montague at the 2002 EnvironMentors Awards Ceremony.

Kevin Coyle (right) and Washington, D.C., director of Toyota Doug West in Anacostia Park, Washington, D.C., in September 2002.

A CONVERSATION WITH
Whitney Montague and Kevin Coyle

Environmental education needs more attention, according to Whitney Montague. "Environmental education helps students discover how the environment affects their community and their family and how they, in turn, can improve the environment around them. The recognition that they are a valuable part of a greater whole opens students' minds to the vast potential within each of them."

Montague is executive director of the EnvironMentors Project at the National Environmental Education and Training Foundation (NEETF), a program that helps students develop environmental awareness. Previously a senior environmental analyst for the Center for Watershed Protection, Montague obtained her B.S. in civil engineering from Howard University. She is currently a featured lecturer and advisor of the Princeton Review Foundation, which provides college preparatory support to economically disadvantaged and minority students.

Kevin Coyle is president of NEETF, which seeks to build a stronger future through innovative environmental learning and improved health, education, business, and ecological protection. Coyle received his B.A. in social work from LaSalle University and his J.D. from Temple University. His work directing NEETF is based on many years of environmental experience. An attorney, Coyle began his career with the U.S. Department of the Interior, supervising the Wild and Scenic Rivers planning program, the Land and Water Conservation Fund, and the Urban Park and Recreation Recovery Program for the northeast region. He also co-founded and was CEO of the American Land Resources Association, and later became president of American Rivers, the nation's principal river conservation organization.

What Is the Issue?

Research shows that Americans know very little about the environment. Most adults can name major environmental problems, but are unaware of what actually causes them or what can be done to prevent them. This is especially true of more complex issues such as climate change, species extinction, or run-off pollution. A study conducted by the National Environmental Education and Training Foundation (NEETF) and Roper ASW shows that two-thirds of Americans fail a basic quiz on the environment and as many as 85 percent lack sufficient knowledge of energy issues. There is even less understanding of basic ecological principles and natural ecosystems such as forests, wetlands, rivers, coasts, and the water cycle. Ecological illiteracy is a large problem.

Environmental education, when done well, can close this gap in public understanding and help people learn more about their direct connection to the environment. Effective environmental education builds an affinity with nature and local ecosystems, especially when educators teach through students' own experiences. It also demonstrates how a healthy environment is an interconnected whole, rather than seemingly separate elements such as a tree, a human, a sea, or a city. Today, despite decades of progress, children and adults are still not receiving enough organized environmental education. This has serious implications for our environmental future and signals an opportunity for environmental careers.

- Untrained Teachers

According to the North American Association of Environmental Education, just 10 percent of teachers have had specific preservice training on teaching environmental concepts and issues. Only one in four has ever taken an environmental course, and only one-third of active teachers have participated in brief training linked to specific courses offering instruction on how to deliver environmental education in the classroom such as those offered by Project WILD, Learning Tree, and WET. Additionally, the 60 percent of teachers who do engage in environmental education each year have low skill levels, which inhibit them from teaching the material successfully.

- Competition Among Coursework

Environmental education is typically carried out for a mere few hours or days each year, mostly as an add-on to core classes. As a subject, the environment is in danger of losing ground because of a greater-than-ever emphasis on statewide learning standards and tests. When students receive low standardized reading and math scores, schools become more interested in raising those test scores (which are the current measure of educational achievement) than in creating interconnected learning or teaching environmental education.

In addition to pressure for obtaining higher test scores, schools often have financial difficulties that draw funding away from environmental education. In most cases, any new money will go toward core courses or textbooks, rather than separate (or integrated, for that matter) environmental education. These schools are less inclined to spend added funds on environmental education because they face pressing needs for basic, core curricula.

- ## High Schools Falling Short

Except for some specialty high schools, environmental education is generally relegated to elementary or middle schools. While more than 80 percent of grade school teachers report they teach about the environment, only half of that percentage cover environmental subjects in high school. Some good news is that a small but growing number of secondary schools now offer more advanced classes in ecological studies and advanced-placement environmental science.

But most high school students are currently underserved. By not receiving well-designed environmental education, they miss out on opportunities to learn about science, mathematics, government, culture, history, and psychology in a uniquely integrated way. The complexities that environment-based curricula throw at students are especially valuable tests of the abstract thinking skills that schools are working so hard to help young people develop. Moreover, these students may also miss the opportunity to discover an environmental career.

- ## The Obstacle of Fieldwork

When teachers actually include environmental education in their lessons, it is usually confined to classroom learning, which is very limiting. One of the most significant opportunities for environmental education is found in the environment itself. Classroom instruction supported by field study is much more effective. Unfortunately, teachers are sometimes reluctant to get students outside.

Field trips, for example, often present significant logistical problems despite numerous local opportunities for students to experience nature even in the most urban school settings. The Chesapeake Bay Outward Bound Foundation's program is a program that ought to have waiting lists. The state of Maryland has contracted the foundation to offer three- to five-day field experiences for middle school and high school students, free of charge. However, chaperones are needed for field trips. Local school systems cannot usually afford to pay for substitute teachers or separate chaperones, and so, all too often the students stay home. This happens in many other fiscally constrained school systems across the country.

• Larger Consequences

The consequences of our environmental education shortfalls can most easily be measured by deficits in skills and knowledge—by the fact that students simply do not learn what they need to know about how nature works and how human activities affect ecological health in this country and throughout the world. Beyond poor test scores in science or geography, poor (or nonexistent) environmental education has even more serious impacts. Students who lack this type of education fail to connect with the natural world and also fail to gain an understanding of themselves as living within a dynamic web of life. Without this understanding, it is likely that these students may grow into the same kinds of citizens and consumers that have helped create so many of our current environmental problems. It is also unfair to deny students the sheer sense of awe, joy, and humility that comes from effective environmental education. Without this integrated curriculum, many students miss out on the fact that learning can be fun when directly applied to everyday life.

How Are Environmental Professionals Approaching the Issue?

Environmental education is not about preaching a pro-environment point of view. It helps students reach their own conclusions and therefore is a tool for building and refining critical thinking. A person with true environmental literacy (as compared to more general awareness) is a person who has enough context and skill to sort through complex environmental issues. If students are incapable of grasping future environmental complexities because they lack this skill and context, there will be many fewer opportunities for solutions. In assessing the public need to build a dam on a river, for example, students must consider wildlife and habitat losses along with positive benefits of the dam. This involves tough balancing. Good thinkers learn to consider many aspects of a problem before deciding what to do, and environmental education helps this process in both environmental and nonenvironmental contexts.

• Developing New School Curricula

Environmental education has become more focused on how it supports the larger education field. To ensure that environmental education does not get bogged down in the "competing subjects" dialogue—or is overlooked altogether in schools—we can use it as an integrated context for learning and incorporate it into ways that we are already teaching. Using this material as a cross-disciplinary subject is extremely effective, because the environment is an integral part of every student's daily experience.

According to the State Education and Environment Roundtable (SEER), integrating environmental education throughout coursework strengthens learning skills and fosters more enthusiastic students and teachers. For example, using environmental literature or statistics in regular writing or mathematical exercises gives context to problems, and in many cases, students' reading and math comprehension then improves. While integrating coursework does not mean that every precalculus class must focus on environment-based problems, some coursework is necessarily and directly connected to the environment. Overall, integrating environmental education into mainstream coursework will help students see important interconnectivity and enhance their overall subject learning. As Dr. Gerald Lieberman, director of SEER, notes: "EIC [Environment as an Integrating Context] holds great promise for helping close the achievement gap in K–12 education."

- Addressing Standardized Testing

Although standardized tests drive the curricula and content taught in many schools, they are arguably not a full and accurate measurement of student learning. To adjust to a greater focus on standards, several environmental

BOX 10-1

STRATEGIES TO INTEGRATE ENVIRONMENTAL EDUCATION

- Developing New School Curricula
- Addressing Standardized Testing
- Helping Teachers to Be Integrators
- Involving Environmental Professionals
- Taking Students Out of the Classroom
- Being Role Models

education groups have matched their programs to state standards in science, language arts, and social studies. They have also documented proof that environment-based programs consistently help improve standardized test performance, therefore giving justification to include this subject in school curricula. For example, SEER research shows that students in environment-based programs performed 27 percent higher in statewide performance assessments than those without environmental education nationwide. Additionally, the environment program in Kentucky correlated with a 10 percent rise in statewide scores in one year.

Despite these successes, environmental education programs require many teachers to risk stepping outside of their normal comfort zone, which is something many teachers are reluctant to do. One way to address this reluctance is to employ organizations, such as Project WILD and Project Learning Tree, that specifically give teachers instruction on how to substitute environmental education for segments of their normal curricula. Perhaps if **environmental educators** were more involved in creating standards to begin with, we might see significant curriculum and testing changes.

• Helping Teachers to Be Integrators

To truly infuse environmental education in schools, we must involve all types of educators and teachers, giving them more incentives to teach environmental material, while stimulating their own enthusiasm and success. School systems could, for example, offer continuing education credits and recognition programs to teachers working on integration. This might set models of behavior for other teachers or entire school systems. Developed at Montana State University, Project Wet is one such program. Project Wet is a K–12 water education program for formal and informal educators. Its exemplary recognition process encourages teachers to participate in environmental education.

Local universities and colleges could also team up with nearby school systems to offer training and use of equipment. University professors and their students might help with curriculum development by providing handouts, materials, guest lecturers, or the equipment necessary for conducting hands-on experiments in school classrooms. If schools offered these services free of charge, many teachers would likely take advantage.

ENVIRONMENTAL EDUCATOR

AT A GLANCE

Employment:

325,000 (including many public school science teachers)

Demand:

Good

Breakdown:

Public sector, 70 percent

Private and nonprofit sectors, 30 percent

Trends:

• Demand for science teachers is strong, especially in urban areas and regions with rising populations.

• Environmental educators are becoming an organized group, as many are taking state certification exams that promote a code of ethics and standards.

Salary:

Entry-level positions earn from $22,000 to $30,000, with median salaries at $40,000 a year. Top salaries can reach up to $80,000 for experienced managers.

JOB DESCRIPTION

Environmental educators focus their efforts on making students aware of environmental issues such as recycling and pollution through classroom learning, hands-on activities, nature walks, and Internet presentations. They strive to provide students with the knowledge and skills they need to tackle environmental issues as individuals or as part of a group.

Many environmental educators say they were drawn to their work in hopes of bringing awareness to youths and adults. Educators who work with children need lots of energy and enthusiasm to motivate young students. They must also convey knowledge in interesting, innovative ways. One of the rewards of this work is the chance to give students "an appreciation, awareness, and sense of stewardship for all life on the planet," says Dot Lamson, director of environmental education at the Chewonki Foundation in Maine.

Educators can work in traditional and nontraditional settings. Science teachers are often responsible for environmental education in K–12 schools, while others work for colleges or universities. Informal environmental education positions, like those at nature centers, aquariums, camps, and zoos, are often seasonal, part-time, or contract jobs. Some educators build a full-time career by serving multiple clients on an independent basis.

Professor Carol M. Tunney, who teaches environmental issues at Southern Vermont College, is a former obstetrician who left the medical field to pursue her passion for the natural world. "I give my students a concrete sense of the most pressing environmental issues today," she says. "My coursework also focuses on the impact these environmental concerns can have on public health. We pay attention to the news from sources like National Public Radio, and we even hold environmental fairs on campus."

GETTING STARTED

• A bachelor's degree in education, environmental studies, or natural sciences is required for employment.

• To teach in a public school, a teaching certificate is necessary. A master's degree in education offers an advantage when looking for a teaching position.

• Student teaching opportunities or summer employment at a camp or nature center are invaluable experiences.

• Some employers, such as camps or nature centers, require emergency training in CPR and first aid.

ADVICE FROM THE PROS

Environmental educators Brett and Laura Holmquist started the Ravenwood Natural Science Center three years ago in Bigfort, Montana. Laura Holmquist is a trained wildlife biologist, while Brett Holmquist is a licensed elementary school teacher. Brett advises that prospective educators "should step into the work part-time before committing to it fully. This is demanding work that requires a lot of energy and constant learning." He also says that determining a specific area of expertise will make you more attractive to employers. "Becoming an expert in one or two areas, such as wildlife or geology, will help you stand out in the crowd."

- Involving Environmental Professionals

The nation has hundreds of thousands of environmental professionals working for public and private agencies, organizations, and companies. They, too, are resources for educators or can become educators themselves. They can help directly by teaching students, training teachers, developing coursework, hosting field trips, and using their considerable expertise in the school setting. For example, religious organizations have recently become much more involved in environmental justice issues in urban areas. Professionals such as **community organizers** at religious and environmental justice organizations have helped form partnerships with schools and have also held their own workshops about imperviousness, rooftop gardens, and idling buses, for instance, to educate about environmental impacts these communities experience. No matter which organizations are involved, any professionals working with schools not only impart environmental knowledge, but also demonstrate job opportunities in environmental fields that many students may not know about.

- Taking Students Out of the Classroom

Environment-based programs that use field study projects, including—but going beyond—field trips, are more successful than those that do not. Research shows that students learn best while doing fieldwork and investigating issues. Combined with classroom instruction, these experiences help build a richer and more durable understanding of the environment and society.

The environmental field as a whole has a vast and underused educational resource at its disposal. Many schools are developing partnerships with local park systems, nature centers, zoos, aquariums, and similar environment-based facilities. Such facilities are increasingly becoming learning centers, placing even more emphasis on environmental education and support for schools and teaching. Nearly 25 percent of the 140 million people who annually visit zoos and aquariums, for example, are there participating in school programs.

Additionally, groups like the Girl Scouts and Boy Scouts that have strong environmental components can provide extracurricular environmental education. Boys' Clubs, Girls' Clubs, nature centers, and recreation centers can take kids on canoe trips or nature hikes, effectively acclimating young students to the environment. They can also provide outdoor activities like Outward Bound programs and community summer camps that teach both experiential environmental learning and critical thinking.

EARTH SCIENTIST

AT A GLANCE

Employment:

30,000

Demand:

Steady

Breakdown:

Public sector, 30 percent

Private sector, 40 percent

Nonprofits, 20 percent

Self-employed, 10 percent

Trends:

- Many professionals will retire within the next decade, creating increased job openings.
- Companies must comply with an increasing number of environmental laws and regulations, creating more geology positions within private firms.

Salary:

Entry-level salaries begin at $35,000, with median annual earnings amounting to $71,725 a year. Most earn between $50,895 and $114,590, while salaries above $145,000 mark the top tier.

JOB DESCRIPTION

Earth science is a diverse and exciting field. Geoscience, geology, geophysics, engineering geology, and environmental geology are just some of the career paths for aspiring earth scientists. Part of the appeal of this work is that basic concepts of geology begin on a global scale. "The history of the site you're investigating, no matter how small, owes a part of itself to some fascinating, gargantuan, geologic event," says Lance Eckhart, an environmental geologist in California.

Earth science as a field focuses on specific aspects of the planet's geological past and present and often makes predictions about its future. Earth scientists spend much of their time identifying and examining rocks, conducting geological surveys, constructing field maps, and studying information collected by remote sensing. Many work for private consulting firms, universities, state and federal government agencies, and oil and gas extraction or mining companies. Scientists working for oil and gas extraction or mining companies often use drilling and seismic technology in their everyday work.

Earth scientists can also work in laboratories using sophisticated equipment to examine the physical and chemical properties of specimens collected during field study. In both laboratory and field work, they use Geographic Information

Systems, computer modeling, data analysis, digital mapping, and geographic positioning systems.

Martha Garcia, program coordinator for ecosystems science at the U.S. Geological Survey, says her research involves studying ecosystems in the San Francisco Bay and Colorado Flat River regions. She focuses on finding a balance between the economic demands for natural resources and maintaining healthy ecosystems. This is sometimes a difficult task, especially when she is under pressure to acquire grants to support her research. "This is a very competitive field," Garcia says. "There's continual need for more advanced research and specialized products and services, which requires us to have state-of-the-art skills in science disciplines."

GETTING STARTED

• A bachelor's degree in geology, geophysics, or another related field is the minimum requirement for employment; a master's degree is preferred.
• Professionals with a master's degree or Ph.D. hold the top research positions.
• Seek internships and jobs that include laboratory work at geological surveys and other federal and state agencies, nonprofit groups, or private firms. Good places to start looking for these positions include the Geological Society of America. (www.geosociety.org) and the American Geological Institute (www.agiweb.org).
• Learn information technology and Geographic Information Systems to help secure employment in this competitive field.

ADVICE FROM THE PROS

Todd LaMaskin, a principal geologist with the URS Consulting firm in Oregon, says students should learn technical writing, statistics, organic chemistry, and, especially, flexibility. "You must be ready for daily challenges because every day is different," he says. "Be willing to accept last-minute changes to your schedule."

Career Spotlight

Project WILD
Houston, Texas

Lynn Garris was teaching chemistry in Alabama's largest middle school when she first heard about Project WILD. "Our principal sent me to a workshop with thirty other educators from around the state. It wasn't until the last day that I realized we were being trained to teach the curriculum to others in addition to our day jobs," Garris recalls. "Years later, I'm now getting paid to do what I'd been doing for free—lead the Project WILD workshops. Life is sweet!"

Her enthusiasm is shared by other Project WILD coordinators, found at conservation agencies in all fifty states and Puerto Rico. Administered by the Council for Environmental Education and co-sponsored by the Western Association of Fish and Wildlife Agencies, Project WILD has been helping K–12 teachers with innovative natural resource curriculums focused on wildlife and habitat for twenty years, and has trained over 900 thousand educators since 1983. Headquartered in Houston, the program's strength lies in its network of coordinators like Garris and its hundreds of trained facilitators who educate teachers at workshops throughout the nation. The ultimate beneficiaries of Project WILD, of course, are the millions of students who learn to understand and appreciate the planet's wildlife and the habitats that support them.

Garris came to her position as a conservation education specialist at the Alabama Department of Conservation and Natural Resources with a B.S. in biology, an M.S. in science education, and several years of classroom experience. Her duties include developing, coordinating, and conducting Project WILD teacher workshops and "train-the-trainer" sessions in southern Alabama.

Her Project WILD counterparts in other states include M. Susan Gilley, wildlife education coordinator in the Wildlife Division of the Virginia Department of Game and Inland Fisheries, and Jeff Rucks, chief of education for the Colorado Division of Wildlife. Not surprisingly, all three note that a shared passion for both nature and children characterizes Project WILD staff people.

Project WILD provides innovative tools and curricula for environmental educators to use in the field.

"I've been concerned about the environment since my Girl Scout days," Gilley says. "During college, I worked as a seasonal naturalist for the National Park Service, and I discovered that I enjoyed passing on my interest in wildlife." For his part, Rucks remembers a Boy Scout childhood backpacking trip in the Sierra Nevada Mountains. His college summers were spent with the U.S. Forest Service as a seasonal firefighter and wildlife technician, where he used his rock climbing skills to rappel down rocky cliffs in search of peregrine falcon nests and his skiing ability to grab a position as a snow ranger in the Rocky Mountains.

Their educational backgrounds are also similar. Gilley has a B.S. in forestry and wildlife resource management from Virginia Polytechnic Institute, as well as a teaching certificate and education experience at a nonprofit math and science center; Rucks earned a B.S. in wildlife management from Humboldt State University and a master's degree in education administration from the University of Phoenix.

Many of the state coordinators have other duties as well. "Besides the Project WILD program, I also supervise hunter education, angler education, 'Watchable Wildlife'—even the 'Becoming an Outdoor Woman' program," Jeff Rucks reports. "Like others in my position, I'm responsible for aspects of K–12 environmental education in the state's schools."

Ensuring that Project WILD fits into, and supports, educational standards from national and state governments and respected institutions —like the National Science Foundation—is an important part of its success, according to Elisa Lewis, senior manager at headquarters in Houston. "To be relevant, we help educators use environmental education as a great method for meeting the government standards that guide so many decisions in our schools," Lewis reports.

In 2000, the program made extensive revisions to its flagship *Project WILD K–12 Curriculum and Activity Guide* and to its companion volume focused on aquatic education. The revised guides incorporated a new environmental education learning framework. In recent years, the program has also developed *Science and Civics: Sustaining Wildlife* and produced a Spanish language version of its educational activities.

Although the Project WILD people are proud of their jobs, they point to K–12 teachers as the real heroes. "Those who can, teach," Elisa Lewis says. "If you're looking for a career that makes a difference, become a teacher. We'll be there to help."

- Being Role Models

Mentors, tutors, and guest lecturers play an important role in informal environmental education. Most environmental professionals entered their field because of direct nature experiences or through the influence of an adult who talked to them about the environment. For this reason, environmental professionals have the opportunity to educate by sharing their enthusiasm and discussing their current work and its importance. Both one-on-one and group mentoring can be effective environmental education strategies that strongly impact young people, particularly middle school and high school students.

The EnvironMentors Project at NEETF is an example of a one-on-one mentoring program. EnvironMentors matches environmental professionals with urban high school students from underresourced communities in order to prepare students for college degree programs and careers in science, technology, and the environment. All EnvironMentors projects require environmental field research, which inherently means students and mentors engage in intensive, hands-on study in the natural world. The program also gives students exposure to environmental careers through its expansive network of current and alumni mentors.

Career Advice from the Expert

The field of environmental education will offer many opportunities over the next ten years. Some 50 percent of teachers working today, for example, will be retiring by 2012. Research on environmental education shows that teachers who use environmental education effectively get good results and instill high levels of learning enthusiasm in their pupils. Prospective teachers and other educators should make it a point to understand more about what environmental education has to offer and become versed in its dynamic teaching and pedagogical strategies. This could be challenging because the current formal education system has yet to recognize the full value of the environment as a tool for education. A little persistence and an open-minded focus on what goes into good teaching will help.

Remember that most teachers who use environmental education in their classes are seasoned veterans who have tried many other approaches. Effective teachers of the future must be able to think and act outside the (classroom) box at a time when there is considerable resistance to innovation. They should concentrate on learning tactics for teaching experientially, fostering investigation, and teaming up with others to work across disciplines. Future teachers will also need sharp integrative skills to work in both classroom and field settings in order to be effective educators and stewards of the environment.

Prospective educators would do well to take basic science or specific environmental science courses, as understanding scientific principles will improve their ability to explain environmental issues to students. It's also important to seize opportunities for practicums that offer exposure to outdoor and nature-based learning labs and venues. This will instill some comfort in working outside of traditional classrooms and foster greater adaptability. It will also show aspiring teachers what it is like to work with students who are jumping up and down with excitement to learn.

Aspiring environmental education job seekers should also be aware of increasing opportunities outside K–12 schools. Zoos, aquariums, science and natural history museums, and nature centers have recently come to recognize their role and value as places to educate people about the environment. As these institutions shift focus from observation and curatorial programs to education, more **environmental educators** will be needed. Tight budgets, however, may restrict the number of new hires.

Informal environmental education centers are also places to seek internships and volunteer assignments that offer exposure to the tools and techniques of environmental learning. **Environmental educators** must be able to explain science and convey mystery. There is no better preparation for becoming an educator than to spend time around those who have been doing it for years.

RESOURCES

Advanced Technology Environmental Education Center, www.ateec.org

Chesapeake Bay Foundation, www.cbf.org

Educational Resources Information Center, www.ericse.org

EnvironMentors Project, www.environmentors.org

National Council for Science and the Environment, www.ncseonline.org

National Environmental Education and Training Foundation, www.neetf.org

National Wildlife Federation, www.nwf.org

North American Association for Environmental Education, www.naaee.org

Project Wet, www.projectwet.org

Project WILD, http://www.projectwild.org

State Education and Environment Roundtable, www.seer.org

11. ENVIRONMENTAL JUSTICE

Dr. Robert Bullard speaking at the 20th anniversary of the Environmental Health Coalition, a toxics watchdog group in San Diego and one of the country's leading environmental justice groups.

A CONVERSATION WITH
Robert Bullard

It would be difficult to find someone more knowledgeable or respected in the environmental justice field than Robert Bullard. He has worked on these issues since 1978—before the term "environmental racism" even had meaning—and his landmark study, *Solid Waste Sites and the Black Houston Community,* led to the first lawsuit that used civil rights laws to challenge environmental discrimination. Since then, he has continued fighting for environmental equity. "Environmental racism isn't going away," he says. "But communities are banding together and, in many cases, winning."

Bullard is currently director of the Environmental Justice Resource Center and the Ware Distinguished Professor of Sociology at Clark Atlanta University. One of the planners of the First National People of Color Environmental Leadership Summit, he served on President Bill Clinton's transition team in the "Natural Resources and Environment Cluster." Bullard is also the author of numerous articles, scholarly papers, and twelve books, including *Dumping in Dixie: Race, Class, and Environmental Quality* (Westview Press, 1990, 1994, 2000), which has become a standard text in the environmental justice field, and *Sprawl City: Race, Politics, and Planning in Atlanta* (Island Press, 2000). He received his Ph.D. from Iowa State University.

What Is the Issue?

Environmental justice embraces the principle that all people have a right to equal protection under environmental, housing, transportation, health, land use, and civil rights laws, and that no one group of people should be singled out for environmental and health hazards.

The environmental justice movement is a much-needed response to the existing environmental protection paradigm in our country, which reinforces the stratification of people according to race, ethnicity, status, place, and power. This framework also manages, regulates, and distributes environmental risks and pollution instead of preventing degradation or providing true health for all people.

These patterns have institutionalized unequal enforcement of environmental laws. Because the government has not prohibited such occurrences, it has effectively traded human health for profits, placed the burden of proof on victims, legitimated human exposure to harmful chemicals, and subsidized environmental destruction. The actions have resulted in promoting "risky" technologies, exploiting disenfranchised communities' economic and political vulnerability, creating an industry around risk management and assessment, delaying cleanup activities, and failing to develop precaution and pollution prevention protocols.

• Where Environmental Injustice Occurs

Low-income neighborhoods, communities of color, and indigenous peoples are groups that are disproportionately affected by environmental hazards. These communities can be located on reservations, in urban ghettos and barrios, or in rural poverty pockets in the southern United States in "colonias" (unincorporated communities), along the U.S.–Mexico border.

Polluting industries tend to "set up shop" where there is a path of least resistance. People living in low-income neighborhoods or communities of color do not usually have the economic or political resources to fend them off. Consequently, these communities are systemically targeted for landfills, incinerators, polluting industries, freeways, and other locally unwanted land uses (LULUs). They also experience unequal protection by and enforcement of environmental laws, resulting in environmental racism, classism, injustice, and degradation.

It's not by accident, for example, that the southern United States is the most polluted region of the country. The entire Gulf Coast region, especially

Mississippi, Alabama, Louisiana, and Texas, has been ravaged by lax regulations and unbridled polluting industries because it's where the poor and people of color live. *Dumping in Dixie* is more than a book I wrote. It documented Louisiana's petrochemical "Cancer Alley" corridor, the area along the Mississippi River between Baton Rouge and New Orleans where cancer rates are sky high due to the chemical refineries that line the river's banks that give off high levels of industrial toxic emissions.

- Land Use Implications

Urban planning and design decisions greatly influence the built environment, largely determining what goes in and what stays out of communities. Probably the most widely applied mechanism to regulate urban land use in the United States, zoning creates "a place for everything, and everything in its place." Theoretically, this should keep a lead smelter from being built next to an elementary school or children's playground, yet such incompatible land uses are widespread in low-income, people-of-color communities.

Nearly half of all public housing in the United States, for example, sits within one mile of a factory that reports toxic emissions to the government. Nearly 1,200 public schools in Massachusetts, New York, New Jersey, Michigan, and California (mostly populated by low-income and people-of-color students) are located within a half-mile of federal Superfund or state-identified contaminated sites. In Camden, New Jersey, polluting facilities operate directly across from a park. In one community after another, we see schools, parks, and playgrounds that are too dangerous for kids because of their proximity to environmentally harmful industries.

- Public Health Impacts

Clearly, where you live can impact your health. In 1999, the Institute of Medicine issued its study, *Toward Environmental Justice: Research, Education, and Health Policy Needs*, which confirmed that low-income and people-of-color communities are disproportionately exposed to higher levels of pollution in their homes, neighborhoods, and workplaces than the rest of the nation. These same populations experience certain diseases, such as cancer, respiratory infections, asthma, chronic obstructive pulmonary disease, and heart disease, in greater numbers.

Environmental hazards such as pollution are not randomly distributed in

231

the United States. The National Argonne Laboratory reports that 57 percent of whites, 65 percent of African Americans, and 80 percent of Hispanics live in the nation's 437 counties with substandard air quality. Ironically, millions of people who live in urban areas do not own cars, although they oftentimes breathe automobile pollution from others who commute to the city from the suburbs.

Pollution generated by motor vehicles has created over 200 nonattainment areas, or areas where air pollution levels persistently exceed national ambient air quality standards. Polluting vehicle emissions contribute significantly to illness, hospitalization, and premature death. Children are at special risk for asthma and other respiratory problems exacerbated by ground level ozone—the main ingredient of smog. For more information about pollution, see chapter 2 on air quality.

Because most metropolitan regions are designed to support suburban sprawl, we continue to mow down trees and build more highways, megamalls, and far-flung housing subdivisions, consequently requiring more use of cars, which means more pollution and possibly more illnesses. Much of metropolitan Atlanta's environmental problems and uneven growth is the result of an "iron triangle" of perverse finance incentives, poor land use planning, and an overreliance of transportation service delivery that rely upon the poor model.

Current housing infrastructure also creates disproportionate public health risks to the denizens of older urban communities. Our country's continued childhood lead poisoning problem, a preventable disease most often found in inner cities where urban or low-income kids live, is probably the most blatant example of this. Recent studies suggest that a young person's lead burden is linked to lower IQ, decreased high school graduation rates, and increased delinquency. Government and private industries, however, do not have the will or commitment to get lead paint out of older housing.

- Economic Repercussions

Transportation equity and environmental justice are intricately linked. Decisions to build highways, expressways, and beltways have far-reaching impacts on economic opportunity, the environment, and public health. Roadway projects have received over $205 billion since 1956. In contrast, public transit has received roughly $50 billion since the creation of the Urban Mass Transit Administration in 1964. This federal agency was created to offer transit assistance by providing $375 million in matching funds to cities and states for large-scale urban public or private rail projects.

The decentralization of employment centers has also had a major role in shaping metropolitan growth patterns and the location of people, housing, and jobs. Government policies have buttressed segregation and suburban sprawl through new roads and highways at the expense of public transit. Tax subsidies made it possible for new suburban employment centers to become dominant outside of cities, pulling middle-income workers and homeowners from the urban core.

• Beyond Our Borders

Most people around the world want jobs and economic development. But at what cost to their health and the environment? No global standards for environmental protection exist, making it easy for transnational corporations to flee to developing countries with the fewest environmental regulations, best tax incentives, cheapest labor, and highest profit. The result is often exploitation. This has also placed special strains (including climate change impacts) on tropical ecosystems in many poor communities and nations inhabited largely by people of color and indigenous peoples. Overall, these special strains are indicative of environmental injustice and further demonstrate the growing gap between the haves and have-nots in the developed and developing worlds.

People all over the world should understand and adhere to certain acceptable practices. Locals working for U.S. companies in Mexico, Bangladesh, or Taiwan, for example, should receive fair wages and protection from environmental and health hazards. If inhaling mercury or lead is hazardous for U.S. workers—so much that they require protective gear—the same chemicals will likely harm workers in other countries. They deserve protection against conditions that we would not tolerate here.

Furthermore, some companies are shipping chemicals, pesticides, and commercial processes abroad that have known carcinogenic, reproductive, or neurological effects. Even if these practices are legal in other countries, they have serious health impacts at home and abroad. For example, some chemicals banned here are used to grow fruits in other countries. The fruit is then shipped back to us, instigating a circle of poison by unjustly exposing both foreign workers and domestic consumers to environmental and health hazards that are outlawed here.

How Are Environmental Professionals Approaching the Issue?

The environmental justice movement attempts to reverse inadequate environmental protection patterns by incorporating prevention, health, and the "Precautionary Principle" into sustainable community design and environments. This redesign asks how little harm is possible, rather than how much harm is allowable, thereby eliminating the threat before harm occurs. Based primarily on a public health model of prevention, this redesign and protection should apply uniformly to communities in Beverly Hills as well as southern Los Angeles.

Environmental justice applies to the places we live, work, play, and go to school, as well as to the greater physical and natural world. It extends the scope of environmentalism beyond its traditional concentration on conservation, wilderness, and outdoor recreation. In this way, it infuses social and political issues with equity and fairness. It also brings together all affected communities as equal stakeholders in order to reflect and respect human diversity.

• Grassroots Organizing

Environmental justice work depends on grassroots, community-based organizations that involve local people in participatory efforts to change local community conditions. Although these groups oftentimes do not have many resources, they usually reside within affected communities and therefore can

---

BOX 11-1

STRATEGIES TO CREATE ENVIRONMENTAL JUSTICE

- Grassroots Organizing
- Educating Communities
- Documenting Injustice
- Fighting Legal Battles
- Planning and Land Use
- Relocating Communities
- Building a Global Movement

---

play a meaningful role in creating solutions. It is their job to mobilize citizens to remedy unfair environmental conditions.

Grassroots organizers have won some major victories over the years. In September 1998, for example, after more than eighteen months of intense grassroots organizing and legal maneuvering, St. James Citizens for Jobs and the Environment forced the Japanese-owned Shintech Inc. to scrap its plan to build a giant polyvinyl chloride (PVC) plant in Convent, Louisiana—a community that is more than 80 percent African American. The Shintech plant would have emitted 600,000 pounds of air pollutants annually.

Educating Communities

Environmental justice is also about educating, training, and informing impacted individuals and communities in a timely, user-friendly manner. **Environmental educators** working with community groups can create educational materials, articles, books, videos, technical reports, training materials, and artwork for public use. The Environmental Justice Resource Center's (EJRC) technical assistance and education program is one successful example. Among its many other activities aimed at facilitating inclusionary environmental decision making, the EJRC provides training for students and community groups on geographic information system analysis, so they can map and develop joint strategies for addressing local environmental problems.

Documenting Injustice

Academic centers like the EJRC or community groups themselves often employ researchers to conduct assessments documenting environmental hazards to prove injustice. These researchers often include toxicologists, who document chemical dangers, or epidemiologists, who determine public health risks. In some cases, the EJRC trains community residents to conduct the research and analysis themselves. By documenting injustice, these researchers help communities speak for themselves with proven authority.

Involving health professionals in environmental justice work is also necessary. Medical doctors and public health advocates can attest that low-income communities and communities of color often suffer disproportionately because they see how injustice contributes to patients' everyday health problems. This testimony and documentation is essential for proving injustice and enabling communities to speak out against discrimination.

COMMUNITY ORGANIZER

AT A GLANCE

Employment:

5,000

Demand:

Remaining constant

Breakdown:

Nonprofits, 100 percent

Trends:

• While environmental activism has tapered off over the last twenty years, it's starting to gain momentum again due to public demand.

• More job opportunities may arise as environmental groups gain support from private companies that are establishing grants and foundations for environmental and social causes.

Salary:

Entry-level salaries begin around $25,000, with the median annual income ranging from $28,000 to $35,000. Top-level salaries reach $60,000 a year.

JOB DESCRIPTION

Community organizers provide citizens concerned about community environmental problems with the know-how to plan successful campaigns and build strong advocacy networks "It's providing a lot of strategy techniques for citizens," says Alyssa Schuren, organizing director for the Toxic Action Center, based in multiple cities throughout the Northeast. "We empower people to win, provide expert advocates like lawyers and scientists, and sometimes even oversee entire campaigns."

Sometimes called campaign organizers, environmental organizers, or outreach coordinators, community organizers work at nonprofit and public interest groups on issues like environmental justice, urban sprawl, conservation, and environmental health. Some organizers work on specific campaign issues, while others simply focus on their organization's mission. Partnership for Parks, for example, concentrates its work solely on the restoration and promotion of New York City's public parks, according to the organization's outreach coordinator Nicole Smith. While Smith's daily responsibilities vary, she says her objective remains the same: to build and strengthen coalitions of people working to conserve and promote a safe and healthy environment, and in this case, New York City's public parks.

Community organizers mainly help citizen groups build alliances with other communities, organizations, and elected officials. They initiate dialogues between these groups to reconcile differences and create media attention to advance their cause. "We introduce people to the whole world of environmental organizing and everything they can do within it," says Leslie Samuelrich, executive director of Green Corps (www.greencorps.org), a group dedicated to providing hands-on training to aspiring environmental activists and community organizers. Often, community organizers must effectively raise money through soliciting and event planning for future activism efforts.

Finally, community organizers often prepare public presentations to educate citizens, facilitate community meetings, and recruit and train volunteers for action. "The challenge is to figure out how to tap into people's concerns about the environment and motivate them to get involved," Schuren says.

GETTING STARTED

• Community organizers come from a variety of backgrounds, but many positions require a bachelor's degree.

• Knowledge of environmental issues is invaluable.

• Employers look for people with strong organizing backgrounds and demonstrated ability in conducting public outreach campaigns. Try to gain volunteer or work experience at nonprofit or student organizations.

• Knowledge of foreign languages can be valuable, especially if working with minorities.

ADVICE FROM THE PROS

Schuren suggests learning about environmental activism at the grassroots level by working for local groups and seeing if the career is a good fit. "These are really hard jobs but also rewarding," she says. "You have to be persistent and willing to attend nightly meetings. This is not a 9 to 5 job, but you're making a difference in people's lives."

- Fighting Legal Battles

Environmental justice is defined as a right, and it's an area where environmental and civil rights laws converge. **Lawyers** file lawsuits and administrative complaints to challenge permits, operating facilities, and pollution. Unfortunately, many disadvantaged and environmentally affected communities do not have money to hire **lawyers** or technical experts. In response, the EJRC created a section in its *People of Color Environmental Groups Directory* that lists **lawyers** and legal organizations that handle environmental justice litigation *pro bono*.

Nonetheless, successful lawsuits have forced our government to become accountable for poor regulation and unequal protection. In 1994, President Clinton issued Executive Order 12898. The order combines Title VI of the Civil Rights Act of 1964, which prohibits using federal funding for discrimination, with the 1969 National Environmental Policy Act, which mandates that all major projects complete environmental impact assessments before building can begin. As federal agencies adopted the executive order, it trickled down to states and municipalities, ensuring that local policies were not disproportionately impacting low-income and minority communities.

While we can celebrate some success of the executive order and subsequent Title VI legal strategies, environmental justice received a legal setback in April 2001, when the U.S. Supreme Court ruled in the *Alexander v. Sandoval* case that communities suffering from disproportionate environmental harm now must prove discriminatory intent—a much higher burden of proof. Clearly, the greater battle for environmental justice is far from over.

- Planning and Land Use

The environmental justice movement as a whole works to integrate transportation, health, environmental quality, economic development, and job accessibility in order to create healthier and better-designed communities.

Urban planners play a particularly significant role in rectifying injustice because they design communities and are involved in planning and zoning that affect where people live and where industries operate. They must collaborate with environmental justice advocates, local grassroots organizations, and educators to incorporate environmental justice principles into their work. It is also crucial to involve leaders of disproportionately affected communities in such collaboration. Then, when **land use planners** develop regional plans or design highways, transit systems, subdivisions, and other facilities, they can

LAWYER

AT A GLANCE

Employment:

55,000 (estimated) of the nation's 725,000 lawyers focused in "environmental" fields

Demand:

Strong

Breakdown:

Public sector, 30 percent

Private and nonprofit sectors, 50 percent

Self-employed, 20 percent

Trends:

• Lawyers encounter fierce competition for jobs, as there are often more law graduates than jobs available.

• State and local government may need environmentally focused lawyers to fill positions created through expected retirements.

Salary:

Starting salaries for lawyers begin at $43,000, with median annual earnings equal to $89,000. Most lawyers earn between $60,000 and $133,000, with the top 10 percent earning more than $150,000.

JOB DESCRIPTION

Environmental lawyers work within the legal system to hold those who commit environmental violations accountable for their actions. They often work to change environmental policy or advocate for a nonprofit group's cause, such as upholding environmental legislation on behalf of citizen organizations. Lawyers can work for governmental agencies, private corporations, or consulting firms, and some are even self-employed. "This interesting field offers opportunities to create meaningful social change," says Delmar Karlen, a lawyer at the Environmental Protection Agency.

Lawyers spend most of their time researching prospective cases, analyzing evidence, filing lawsuits, writing briefs, and meeting with clients and witnesses. Since some environmental law involves litigation, lawyers can spend time in court arguing their cases. Courtroom lawyers must be able to think and speak quickly, easily, and with authority. When cases do not go to trial, lawyers meet with their opponents in order to work out a settlement.

In and out of the courtroom, lawyers need good mediation skills, as their primary goal is bringing parties together over controversial issues, such as land use or pollution control. Lawyers are also known for being skilled negotiators, experts in conflict resolution, and insightful analysts. They sometimes serve as consultants and advise their employers on courses of action that adhere to environmental law.

GETTING STARTED

• A bachelor's degree is required for admission to law school. While no specific undergraduate major is a prerequisite, coursework in environmental studies is helpful.

• Aspiring lawyers must attend a law school accredited by the American Bar Association (www.aba.org) in order to qualify for the bar exam. Look for law schools that have strong environmental programs.

• Lawyers must pass the bar exam and earn a jurist doctorate before they enter public or private practice.

• Look for internships or work experience that will introduce you to litigation and courtroom procedures at either private law firms or as a clerk for a judge.

ADVICE FROM THE PROS

Stephanie Kaplan, a self-employed lawyer in Vermont, works with grassroots and citizens group that advocate for environmental issues in New England and abroad. "I help people understand the legal process, which can be daunting," says Kaplan. "We do win sometimes, but it can be frustrating." She also says that prospective environmental lawyers must have a strong stomach and continued optimism to survive in this field. "Environmental causes are often looked at with disdain. You're not going to become rich in environmental law, but at least every once and a while you can stand back and say that you're doing something you believe in."

recognize past land uses that were partially responsible for environmental injustice and rectify those mistakes.

Local community members who become more aware, informed, educated, and involved in environmental justice campaigns often go on to assume professional positions where they have the power to defend the integrity of their neighborhoods. In particular, we have seen an increasing number of low-income citizens and people of color elected to public office or appointed to zoning boards and commissions, where they can advocate balanced planning.

- ## Relocating Communities

While some communities have been victorious in fighting injustice by employing the strategies discussed above, many have not. When all else fails, relocating communities that are environmentally impacted may be the only way for them to live free from environmental harm. In recent years, several communities have been relocated because they could not be properly cleaned up. In April 2001, for example, a group of 1,500 neighborhood plaintiffs in Anniston, Alabama, were relocated because of contamination from PCBs from a nearby chemical company owned by the powerful agricultural corporation Monsanto. Two years later, Monsanto Company, Solutia Inc., and Pharmacia agreed to pay $700 million to settle two lawsuits brought against them by 20,000 Anniston plaintiffs alleging damages from PCB contamination. Relocation, however, is a last-resort strategy. If environmental hazards can be prevented in the first place, people will not be forced from their homes.

- ## Building a Global Movement

We must all come together to fight environmental injustice and promote healthy and sustainable communities. Only with shared understanding and greater collaboration can we truly promote equity, justice, and environmental protection for all people. The environmental justice movement does not end at town, state, or national borders. Rather, it has taken steam on a global level.

In 1991, the first National People of Color Environmental Leadership Summit was held in Washington, D.C. A conference planned for 300 people had over 1,000 attendees, including representatives from every state and a dozen foreign countries. The 2002 follow-up summit attracted over 1,400 attendees. The success of the first summit led to the drafting of the "17 Principles of Environmental Justice," which were translated into a half-dozen lan-

Career Spotlight

Alternatives for Community and Environment (ACE)
Boston, Massachusetts

Although small—with a staff of just ten people—Alternatives for Community and Environment (ACE) has been achieving victories in collaboration with low-income communities and communities of color in the Boston metropolitan area since 1994. Along the way, it has secured a reputation as one of the nation's leading environmental justice groups.

Executive director Penn Loh outlines the challenge ACE faces. "Neighborhoods like Roxbury bear the brunt of environmental injustices in this region," he reports. "We have abandoned and contaminated properties. We have the poorest air quality and the highest asthma rates. We have waste sites that wealthier communities would never accept or would have cleaned up years ago. Even though we need public transportation the most, we see services declining and prices rising."

Loh came to ACE in 1995 with an M.S. in environmental science and policy from the Energy and Resources Group at the University of California–Berkeley, and a background as a senior researcher with two respected progressive think tanks—the Pacific Institute in Oakland and the Tellus Institute in Boston. At ACE, he quickly rose from research director to associate director and finally to his current position.

As executive director, Loh manages the organization, raises funds, and often serves as the point person for lobbying and collaboration with other organizations and agencies. Current ACE programs include a "Roxbury Safety Net" initiative organizing public housing, legal, and technical assistance; management of the Greater Boston Environmental Justice Network, an innovative initiative for transportation justice that includes the creation of a Transit Riders Union (TRU); and REEP, the Roxbury Environmental Empowerment Project.

REEP's leadership program for neighbor-

ACE Executive Director Penn Loh educates students from Timilty Middle School about water quality at the 2001 REEP graduation. Source: Alternatives for Community and Environment.

hood young people is central to the ACE vision of community involvement and empowerment. And Klare Allen and Jodi Sugerman-Brozan are central to this program.

Allen and Sugerman-Brozan arrived at their positions as program co-directors with strong activist and education credentials—one with a community organizing focus and the other from a policy direction. Well before coming to ACE in 1995, Allen was an experienced organizer and founder/director of a coalition supporting homeless women and their children. Her teaching ability has since won her a Conservation Teacher of the Year designation from the Massachusetts Audubon Society. Sugerman-Brozan brings an M.A. in urban and environmental policy from Tufts University and previous experience as the policy director for Save the Harbor/Save the Bay in Boston, where she also created a successful youth education program.

If the backgrounds of Loh, Allen, and Sugerman-Brozan are representative of the research, organizing, policy, and education aspects of ACE, Quita Sullivan and Eugene Benson are the faces of a fifth capacity—legal assistance.

"Just working to ensure fair enforcement of existing laws could be a full-time job for a big team of lawyers," says Sullivan. "With such limited resources, we have to be creative." The ACE attorneys provide representation directly, partner with community groups, and help coordinate the Massachusetts Environmental Justice Assistance Network, a group of more than 100 attorneys, health professionals, and environmental consultants who provide *pro bono* assistance.

Both Sullivan and Benson have worked as attorneys in this field for a few years now. Before taking the ACE position, Sullivan earned her law degree from Wayne State University in 1997 and worked as the environmental justice staff attorney at the Sugar Law Center in Detroit. Benson joined ACE in 2003 after serving as the associate general counsel (leading the Environmental and Regulatory Law section) for the Massachusetts Water Resources Authority. He received his law degree from Georgetown University in Washington, D.C.

Whether through organizing, legal assistance, research, policy, lobbying, or education, ACE staff members consistently strike a note that they feel differentiates activist work from environmental careers in government, business, or academia—a sense that community empowerment is a critical goal.

"We want to win battles, of course", says Allen, ACE's community organizer. "We want the air to get cleaner and the brownfields converted. But we don't only sit down with other professionals and hammer out deals. It's the people in the neighborhoods that are doing this work. Every campaign needs to make the community a little stronger."

guages and incorporated into discussions at the 1992 Rio Earth Summit. Ten years later, at the World Summit on Sustainable Development, held in Johannesburg, the concept, definition, and language of environmental justice were pervasive.

The concept has now taken root in different parts of the world, including developing nations in the Caribbean, South America, Africa, and Asia. This is particularly important for holding transnational corporations accountable, as the globalization of trade and commerce has oftentimes resulted in the exploitation of disadvantaged communities abroad. In this regard, international environmental justice does more than protect our own backyard, and more than merely prohibit the shipping of pollution, toxic products, and injustice abroad. Rather, it ensures understanding and responsibility for our own consumption and behaviors on a global scale. There are still millions of people, however, who are marginalized and live in degraded areas. It is up to the global movement to continue this imperative work.

Career Advice from the Expert

Because of the movement's inclusionary definition of the environment, few disciplines or specializations are excluded from this work. Since there is enough work to go around, it's our goal to have all hands on deck. I tell my students that social scientists can provide historical context for environmental justice problems. Economists can compute a price tag on the remedies. Geographers can map the locations of environmental hazards. Health scientists can assess risks and impacts. **Lawyers** can seek legal redress. **Journalists** can expose problems through investigative reporting. And environmentalist and human rights groups can place the problems on the global agenda for action.

No matter what career path one has chosen—whether in government, private industry, academia, law, civil rights, or environmental or faith-based organizations—aspiring professionals should seek to advance environmental justice in their daily work. Every little bit helps.

My advice to people wanting to work effectively in this field is to hone your skills in respective disciplines and always be a student at heart, willing to listen, learn, and explore new ideas. Be prepared for setbacks, but also take time to celebrate your victories. We do win from time to time. Do not celebrate too long, though, because the issues keep coming back. The key is knowing when you have won and understanding that winning occurs at different levels.

Today, dozens of book publishers are looking for fresh ideas on environmental justice. Aspiring academics should know that it is possible to get promotions, tenure, and merit pay raises in this field. My advice to those people is read, write, and publish, publish, publish.

Finally, information is power. Resource gaps between the haves and have-nots make our job difficult, but not impossible. We must remember that the digital divide is real. It's a special problem in low-income communities of color and more acutely in developing nations—the same populations that are disproportionately impacted by environmental problems. To remedy this situation, prospective professionals should focus their environmental justice work on leveling the playing field by bringing information to the table through participatory and community-driven research, community epidemiology, community-university partnerships, and train-the-trainer education models.

We have made tremendous strides over the past two decades, but much remains to be done to erase the ugly stain of environmental racism. The future of the environmental justice movement rests with young people.

RESOURCES

Alternatives for Community and Environment, http://www.ace-ej.org
Asian Pacific Environmental Network, http://www.apen4ej.org/
Center for Community Action and Environmental Justice, www.ccaej.org
Center for Health, Environment and Justice, www.chej.org
Center for Race, Poverty and the Environment, www.crlaf.org/crpe.htm
Environmental Background Information Center, www.ebic.org
Environmental Justice Resources Center, http://www.ejrc.cau.edu
Indigenous Environmental Network, http://www.ienearth.org/
National Black Environmental Justice Network, http://www.ejrc.cau.edu/
 inbeejccpage.htm
Thurgood Marshall Environmental Justice Legal Clinic, http://www.tsulaw.edu/
 environ/enviropg.htm
Urban Habitat Program, http://www.igc.apc.org/uhp/

12. FISHERIES

Carl Safina with a
leatherback sea turtle.

A CONVERSATION WITH
Carl Safina

"It's OK to use the sea," says Carl Safina, "but not to use it up."
Safina is one of the original and most prominent voices sounding
an alarm that the abundance and vitality of life within the world's
oceans is seriously compromised. He is currently president of the
Blue Ocean Institute, a nonprofit organization in Cold Spring Har-
bor, New York, that is dedicated to inspiring a wider, more cultural atmosphere for ocean conservation through science, art, and
literature. He has lived close to the sea all of his life, fishing since
childhood.

Safina was previously vice president for Ocean Conservation at
the National Audubon Society, where he founded its former Living Oceans program. He received his Ph.D. in ecology from Rutgers University. He is author of more than one hundred publications, including *Song for the Blue Ocean* (Henry Holt Co., 1998)
and *Eye of the Albatross: Visions of Hope and Survival* (Henry
Holt Co., 2002), and he is co-author of the *Seafood Lover's
Almanac* (National Audubon Society, 2000).

What Is the Issue?

My lifetime near and on the ocean and my travels around the world have shown me how widespread the problem of overfishing is. The wild animals people are most interested in eating, such as cod, tuna, grouper, shark, swordfish, and snapper, are at or near their lowest levels in history in much of their ranges. Analysis by fisheries biologists Ransom Myers and Boris Worm indicate that the abundance of many of the large fish species that fishermen are targeting has declined an average of 90 percent. In only a few places have people tried to let fish recover. When they have, some fish have made striking comebacks.

However, the changes in ocean fish populations make ocean ecosystems different than ever before. This has major consequences for both humans and nonhumans. When a group of algae-eating fish is removed from a reef, for example, algae may increase enough to smother corals, ultimately destroying the coral reef system. Similarly, in some places, clams and oysters—which filter water as they feed—have been depleted to such low densities that water quality has seriously deteriorated. Many other examples exist.

• Overfishing

Overfishing is the modern-day buffalo hunt. There were once 60 million buffalo in North America, which humans drove to near-extinction. Similarly, commercial hunters believed they could never affect the multitudes of passenger pigeons, once the most abundant bird in North America; they drove this bird to extinction in 1914. Many fisheries also were believed to be inexhaustible. Few realized until the 1970s that we were taking out more fish than the oceans could naturally produce. This problem's widespread severity was not generally appreciated until the late 1990s. Now, severe fish depletion is an ecological calamity, threatening human needs. The world fish catch is currently around 100 million tons annually, but overfishing and depletion have produced generally declining catches since the early 1990s.

The way we fish has changed. In the past, people caught fish with primitive traps, nets, or lines. Today, people use large boats with satellite navigation, miles-long driftnets, bottom trawls (nets dragged along the seafloor), and long lines up to eighty miles long with thousands of hooks. In the tropics, many small boats commonly fish with cyanide poison and dynamite or other explosives.

We also fish very wastefully. Approximately one quarter of all sea life taken from the ocean is unwanted and shoveled overboard, dead. Such "bycatch" is the largest source of mortality for adult sea turtles around the world and is also causing declines in albatrosses, many species of fish, and other sea life.

- Population Increase and Pressure to Feed the World

Increasing human population means that more people are eating their way through wild animal populations in the oceans (as well as on land). When a species becomes functionally extinct from overfishing, fishers move on to another species and then another. This "serial depletion" causes fisheries' collapse. Although the need to feed an increasing population is not to be ignored, in the case of fisheries' decline, the larger issue is how to protect the world from the increasing appetite of billions of people.

- Aquaculture

Aquaculture, or fish farming, is often seen as a strategy to counteract fisheries' decline and feed increasing populations, but it too poses various problems. Fish farmers often destroy natural habitats like mangroves or wetlands to create fish farms. And most fish and shrimp eat other fish, meaning that aquaculture still demands wild-caught fish for feed. Before sale, a ten-pound farmed salmon might have been fed thirty pounds of wild-caught fish from the ocean. So aquaculture often uses more fish than it produces. Although farmers are gaining profits, fish farming in this way reduces food sources and fish populations.

Aquaculture can only remain compatible with conserving natural systems by focusing on net protein producers, such as vegetarian species like tilapia or carp, and on small-scale operations that do not significantly damage habitat, such as growing naturally feeding shellfish in floating cages. Unfortunately, these options are not utilized enough.

- Local and Global Effects of Fisheries' Depletion

Overall, depletion has serious social and economic impacts on local communities that depend on fishing. The fishing industry is made up of fishers, fish handlers, boat builders, and all other supporting and distributing services. For the first time in history, coastal communities dependent on fishing cannot sustain themselves, as evidenced by fishing communities' dissipation in eastern Canada and New England. In these areas, people who grew up in fishing families cannot go into this industry.

At the same time, corporate fishing ships are moving all over the world. After fishers have exhausted one area, they have moved to other areas such as the coasts of Africa and some of the South Pacific islands, taking fish away from local people. Industrialized countries also pay for imported fish. Distant markets give people in developing countries incentives to deplete their local waters, putting tremendous strain on local resources.

- Ineffective Government

Although governments have banned some fishing practices, such as using driftnets longer than 2.5 kilometers, current laws and regulations are inadequate. Additionally, government subsidies continue to support overfishing. Mostly, government has been a service agent to the fishing industry rather than a resource defender and fishing regulator. Poor regulation, lack of enforcement, and governments acting as developers rather than resource stewards are conspiring forces that have contributed to reduced fish populations.

- Pollution

The general public usually thinks the main threat to oceans and fisheries is large oil spills. This assumption is incorrect, although those spills do have devastating and long-term effects at the local level, like the 1989 Exxon-Valdez oil spill, for example, where many toxics still affect wildlife today. Fertilizer and sewage runoff are also major sources of pollution, as are air pollutants that eventually fall into the water. For instance, mercury emitted into the air from coal-fired power plants ultimately enters aquatic systems. In these ways, toxics and dangerous estrogen-mimicking chemicals can enter our waterways, where they then affect animal reproductive behavior.

- Global Warming

Fish populations are inevitably affected by serious environmental problems other than pollution, such as global warming, which is resulting at least in part through human activities. Global warming is causing tropical waters to overheat and polar ice to melt, affecting coral reefs and ice-dependent animals. Tropical temperature spikes are causing corals to "bleach," or lose their symbiotic algae, which can kill the corals altogether. In the North Atlantic, melting ice is cooling waters like ice cubes in a drink. Among other things, such changes in water temperatures appear to be retarding the recovery of cod populations. Still other problems include ozone thinning and increased ultraviolet radiation, which can be harmful or fatal to certain kinds of plankton that are fundamental to life in the ocean. For instance, krill, which form the basis of the Antarctic food chain, must find food by grazing the underside of ice when young. As ice shrinks, krill are losing habitat, causing food shortages for penguins and other Antarctic krill-dependent wildlife.

- Nonnative Species

Humans introducing nonnative species can also devastate local wildlife. Nonnative species primarily include fishes, disease organisms, snails, plants, and invertebrates. People introduce alien species deliberately to create new resources for food, aquaculture, or sport fishing purposes. Also, species are introduced accidentally in ships' ballast water. No matter how they enter natural ecosystems, introduced species can cause disease among local populations or outcompete native species, seriously disrupting normal ecosystem functions. The introduction of green crabs, for example, which prey voraciously on local shellfish, has altered these populations on parts of the United States east coast.

- Habitat Destruction

Human activities are also directly destroying fish habitats. Approximately half of the human population lives within seventy-five miles of the coast. Development of coastal areas often destroys fish habitats such as wetlands and mangroves, which are important fish-spawning areas and nurseries. To cite one example among thousands, development has destroyed 90 percent of California's wetlands. For more information about coastal development and its impact on marine habitat, see chapter 6.

How Are Environmental Professionals Approaching the Issue?

- Establishing Protected Marine Reserves

Many people believe that closing some areas to fishing by creating "no-take marine reserves" will be an effective way to manage ecosystems and replenish fish populations. Linked networks of reserves could help sustain full, healthy ecosystems. Scientific literature shows that inside protected areas, fish grow larger, more abundant, and usually more diverse. Other members of the living community likewise benefit, as reserves prohibit dredging, dumping, construction, and other direct disturbances. It is less clear, however, whether these advantages will guarantee benefits to people fishing outside reserves through a "spillover" of fish that may wander past protected-area boundaries. In some cases this effect has been observed; in others it has not. Fisheries biologists must conduct more detailed research in this area.

Although reserves have been successfully implemented in various countries, including the Philippines, New Zealand, Tanzania, St. Lucia, and many others, few have been established in the United States. Fishing groups often oppose reserves because they are not convinced that reserves will eventually help their fishing interests. Unfortunately, the American conservation community has alienated the fishing community on this issue by insisting that all reserves be closed to any fishing. The reserve concept would likely be more effectively promoted as part of a comprehensive zoning strategy. This strategy would involve designing a range of protected and multiple-use areas and balancing closed areas with zones designated for commercial and recreational fishers' guaranteed access.

Professionals involved in protecting habitats, creating integrated ecosystem management plans, and setting up marine reserves include **marine scientists**, who work to understand the biological dynamics of these systems; **information technology specialists**, who manage the large volumes of data on marine environments; **policy analysts** and **lobbyists,** who use scientific information to convince government agencies to implement certain strategies; local communities and **environmental protection technicians**, who encourage such practices; and **environmental regulators**, who enforce and ensure that fishers do not

BOX 12-1

STRATEGIES TO PRESERVE FISHERIES

- Establishing Protected Marine Reserves
- Creating and Enforcing Laws to Change Fishing Practices
- Designing Eco-Certification Schemes

INFORMATION TECHNOLOGY SPECIALIST

AT A GLANCE

Employment:

358,000

Demand:

4.8 percent growth annually 2004–07

Breakdown:

Public sector, 45 percent

Private and nonprofit sectors, 55 percent

Trends:

- An explosion of new technology is creating more job opportunities for information technologists.
- Many upper-level IT managers are retiring or moving into other fields, opening more jobs for entry- and middle-level employment.

Salary:

Entry-level salaries for managers are approximately $50,000, with median annual incomes from $62,500 to $106,000. The top 10 percent earns more than $130,000.

JOB DESCRIPTION

Information is the key to successful planning and decision-making in the environmental world. Policy makers, scientists, planners, and many others use information and data drawn from myriad sources to help them perform their daily work.

Information technology (IT) specialists are the experts who gather this data by using and designing technological systems. The immense accumulation of biological and geophysical information is useless, however, if it cannot be managed and presented in such a way that it can be practically applied to real conservation problems.

Environmentally-focused IT professionals meet the challenge by combining data from remote sensing, Geographic Information Systems (GIS) and Global Positioning Systems (GPS) into more traditional database structures. Some programmers or systems analysts devote their careers to using technology for very specific causes. "I'm particularly interested in how digital technologies can expand access to information and urgent social and environmental problems," explains Bill LaRocque, director of information services at the World Resources Institute, who is among those professionals balancing technological and issue-oriented interests.

253

On a daily basis, IT specialists spend their time working on database maintenance, data researching, programming, mapping services, and compiling and editing information gathered in research. They then make this information readily available to environmental professionals such as wildlife biologists, geologists, and engineers. This work results in concise reports, allowing readers to grasp the information as it applies to the research. The need for accessible and streamlined data is so important that the United Nations factored it into their 2002 Global Environment Outlook priority list.

GETTING STARTED

• The best job opportunities exist for applicants with a master's degree in computer science, engineering, geography, cartography, or GIS and GPS.
• Learning specific programs used within the larger systems, such as ArcView within GIS or VisualFlight within GPS, will give you an advantage.
• Stay abreast of current technology and trends by attending training sessions like the Urban and Regional Information Systems Association (www.urisa.org).
• The Sloan Career Cornerstone Center (http://careercornerstone.org) also provides IT career and job information.

ADVICE FROM THE PROS

David Fox, GIS coordinator for the U.S. Army Corps of Engineers, specializes in information technology work at the corp's Seattle bureau. In this capacity, he provides research and GIS information to environmental scientists. "Everything we do has an environmental tie to it, so it's important that you have some understanding of the natural sciences," he says. "It also helped me to learn the social sciences to gain an understanding of why we have these environmental problems. In any case, make sure you have a solid IT background because you need every technological tool to evaluate environmental data."

enter marine reserves. Fishers must also participate, cooperate, and support these efforts; otherwise they will be difficult, contentious, or futile in most places.

- Creating and Enforcing Laws to Change Fishing Practices

Current regulations and practices have often proved inadequate in protecting our ocean wildlife populations. In many cases, fishing has driven marine wildlife to all-time lows. Our laws should include bans on all fishing technology that significantly harms natural habitats. For example, bottom trawls should be prohibited at least on hard bottoms and places where corals and sponges grow, as they are most damaging to these environments. Fishing with poison and explosives should be banned as well, but in many areas where they're already banned enforcement is lacking.

Marine scientists and fishery managers should set laws and strict limits on how many fish can be caught, based on species' life histories and population levels. We must also allow fish to lay enough eggs or produce enough offspring to keep a population viable. In many cases, this seemingly obvious conclusion is completely ignored, as fishers often directly target juvenile fish before they've had a chance to reproduce, or target adults in the act of congregating for breeding. Fishery managers will need, in many cases, to prohibit such practices, in order to effectively foster fisheries' recovery and repopulation.

The few success stories of recovery usually result from regulations designed to reduce fishing pressure and allow increased spawning. In the small number of places where fishers follow tight regulations, fish populations are doing well. For example, Alaska's Fisheries Management Council has focused on protecting fish from depletion, so that fishers may make a living off the "interest" of abundant populations rather than depleting the capital.

By adhering to regulations and fishery protection, commercial fishers can play a positive role in preventing further decline, prompting recovery, and rehabilitating collapsed fisheries. When people protect their fishing, fishing deteriorates; when they protect their fish, fishing improves. Although it's too late to prevent collapses that have already happened, shifting the focus to recovery can create larger, healthier populations in the future.

- Designing Eco-Certification Schemes

Eco-certification, or "green labeling," can help consumers create demand for better fishing practices. Green labeling refers to independent evaluation based

ENVIRONMENTAL JOURNALIST

AT A GLANCE

Employment:

4,000

Demand:

Slow

Breakdown:

Private sector, 100 percent

Trends:

- Most environmental journalists work several years for smaller companies on general assignment or other beats before they are promoted to their specialized positions at larger media outlets.
- Media time and space devoted to environmental and sustainability concerns is unfortunately not increasing.

Salary:

Entry-level salaries begin at $30,000, with the median annual income ranging from $35,000 to $50,000. Top salaries reach $60,000 or more.

JOB DESCRIPTION

Environmental journalists face many challenges in their work, as they bring a wide variety of concerns and issues to the public's attention. At times, journalists will act as educators, defining a discovery or explaining the advice of experts. "Environmental journalists have an especially challenging beat, as it is wide-ranging in scope," says Beth Parke, executive director of the Society of Environmental Journalists. "Reporters in this complex area must become experts in identifying stories and communicating their importance to editors and specific audiences."

Journalists may focus on issues including marine fisheries, land conservation, development, pollution, and a variety of scientific issues in order to inform the general public. "If the public is to help create well-informed environmental policy, it needs solid information on which to base its values, choices, and modes of action," says Wendy Redal, of the Center for Environmental Journalism at the University of Colorado. "That can only be provided through dedicated, responsible journalism about the environment."

Like others in the field, environmental journalists investigate leads and tips, research their topics, analyze documents, observe events at the scene of a story, interview witnesses and experts, and report on late-breaking news. They must be resourceful in finding information and persistent in acquiring sources for their stories. Daily deadlines can sometimes make their work hectic and stressful, although some stories may take longer to produce.

GETTING STARTED

- Many employers require a bachelor's degree in journalism, while others look for job candidates with specific majors in environmental studies, science, public policy, or even political science. For television reporters, it is important to attend a college with a strong broadcast major.
- Most print, broadcast, and online media companies look for journalists with three to five years of reporting and writing experience.
- Look for entry-level opportunities at small weekly or daily newspapers after graduation. Broadcast or online journalists also usually start at smaller companies.

ADVICE FROM THE PROS

Dan Fagan, an environmental journalist for *Newsday*, began his career as editor-in-chief of his college paper, then as a general assignment reporter for a small daily newspaper in Sarasota, Florida. He spent a few years working for *Newsday* as a political reporter before landing his environmental beat. "Differentiate yourself from the rest of the pack by taking as many science courses as possible in college so you're familiar with environmental topics," says Fagan. "Get as much journalism experience as you possibly can. Go to the best media organization that offers you a job, no matter where it is."

Career Spotlight

National Fisheries Program
United States Fish and Wildlife Service
Northeast Region

The mission of the National Fisheries Program of the U.S. Fish and Wildlife Service (USFWS) is to conserve, protect, and enhance fish and their habitats for the benefit of present and future generations. To do this, the service provides funds to support state and local fisheries programs; enforces federal laws protecting habitat; carries out science projects; and develops partnerships with conservationists, anglers, and others who care about the nation's fish populations.

In the Northeast region, headquartered in Hadley, Massachusetts, twenty-five fishery management offices and national fish hatcheries work to protect dozens of species. Atlantic salmon, striped bass, American shad, river herring, sturgeon, horseshoe crab, American eel, and lake trout are just a few of the better known creatures the eastern region staff protects—both for the health of the fish themselves and because stable fish populations are important for healthy river systems.

The National Fish Hatchery System (NFHS) has a major and unique responsibility in the USFWS efforts. NFHS helps to recover species listed under the Endangered Species Act, restore native aquatic populations, mitigate fisheries lost as a result of federal water projects, and provide fish to benefit Native American tribes. NFHS employees work closely with other service biologists and with the states, tribes, and the private sector to complement habitat restoration and other resource management strategies for maintaining healthy ecosystems that support healthy fisheries.

Michael Odom and Stephen Jackson are two of the region's fisheries scientists. Odom began working with the Department of Natural Resources as an assistant to a district fisheries biologist. "That experience got me 'hooked' on pursuing a career in fisheries," he says with a smile. With a B.S. in wildlife biology from Colorado State University and an M.S. in fisheries from Virginia Tech, he is now the project leader for Harrison Lake National Fish Hatchery in eastern Virginia. Currently, he supervises staff and station activities, rearing and releasing three to five million larval shad, striped bass, and blueback herring; he also assists management biologists in field surveys monitoring Virginia's rivers.

Fishery biologist Stephen Jackson takes on the daily duties of rearing fish at a national hatchery in Gloucester, Virginia. Jackson joined the service after graduating with a B.S. in marine sciences from the U.S. Coast Guard Academy and serving in the Coast Guard. He describes his work as caring for eggs, fry, and fish during spawning, growing, and distribution, and monitoring water quality and the health of the fish. "If we do our job right, we may help bring their populations back to where they need to be," says Jackson.

Central to the entire Northeast region's fisheries responsibilities is the health of the Chesapeake Bay/Susquehanna River watershed—a basin of 64,000 square miles that drains parts of Delaware, Maryland, Pennsylvania, New York, Virginia, West Virginia, and the District of Columbia. The Chesapeake Bay is the largest estuary in the United States. Its watershed contains an array of habitats, including mixed hardwood forests of the Appalachian mountains, grasslands and fields of agricultural areas, lakes, rivers and streams, wetlands and shallow water habitats, open water of tidal rivers, and more. These habitats support thousands of different species of fish and wildlife, including thirty-nine federally listed threatened or endangered species.

Lisa Moss is both concerned and hopeful about the task at hand. With a B.A. in environmental science from the University of Maryland Eastern Shore, Moss is a fishery biologist. She started her career with USFWS through the Student Career Experience Program (SCEP) and gained firsthand experience with hatchery management involving the successful propagation of imperiled and endangered fish species. Today, she conducts assessment and monitoring of native and nonnative fish and aquatic species, and works throughout Virginia to protect often-threatened fisheries. "It's great work," she says, "operating a twenty-five-foot Boston Whaler, interpreting data collected, writing reports. I have learned myriad skills and found that a really resourceful approach is necessary."

With only 800 professionals, the fisheries program staff simply cannot reach its goals alone. "Partnerships with local government, scientists, and anyone who cares will make the difference," Moss says. "Building those partnerships and making them work is the main task ahead."

U.S. Fish and Wildlife Service biologist Lisa Moss (left) and colleague identify and measure fish at Dyke Marsh on Virginia's Potomac River.

on certain standards. **Environmental certification specialists** and sometimes auditors are responsible for evaluating products and issuing on-pack logos such as the "Dolphin-Safe" label for those operations that comply with environmental standards.

Trustworthy green labels can help consumers make responsible purchasing decisions. When buying seafood, people can vote with their wallets to shift fish consumption and the market toward those species caught sustainably. In this way, eco-certification fosters increasing consumer demand that can also help shift the fishing industry's internal sense of competition, thereby encouraging best practices. The Marine Stewardship Council is an organization that specifically exists to certify seafood products. Their label has helped consumers buy more responsibly, and also created market rewards and incentives for more responsible fishing practices.

Another popular approach for helping consumers make informed choices are the various seafood rating guides published by groups including Environmental Defense, Monterey Bay Aquarium, and Blue Ocean Institute. These groups evaluate and publish lists, guides, and books that rate seafood based on an evaluation of species' population trends, habitat considerations, bycatch, management, farming practices, and reproductive potential, all of which influence consumer purchasing decisions and therefore fishing operations.

Career Advice from the Expert

There is a wide set of avenues for people to make contributions in ocean conservation. Aspiring professionals should seek the highest and best training available in the areas that interest them. Someone interested in nature and natural populations might obtain a high-level degree in ecology; someone interested in the political framework that provides regulations and effective changes should look for an advanced degree in policy or law; someone interested in film or screenwriting can learn those skills as ways toward educating and inspiring the public about the power and magnificence of the seas and how the oceans are changing.

People of all disciplines who work in this area must have a basic understanding of ocean science and the larger context. Prospective fisheries professionals should have an academic background in environmental science, marine biology, or ecology. Strong communications, writing, and speaking skills are becoming increasingly important for anyone interested in conservation, because of the need to galvanize the public and decision makers.

Simply going fishing can perhaps most importantly help prospective professionals gain perspective on the ocean. Surprisingly, many people working on these issues have little understanding of what fishing is about and how the fishing industry operates. Volunteering with research teams or conservation projects or doing a stint on a commercial fishing boat are valuable experiences that can help people entering this field. Some fisheries professionals occasionally go to sea for extended periods, and familiarity with the experience helps.

An excellent strategy for obtaining the right experience and education is reading professional literature to identify the top creative people whose scientific, artistic, or policy work most excites you, and then volunteering or arranging to work or study with them. One word of warning, however. Volunteer internships and research assistantships are often tickets to adventure and valuable experiences that can start a career, but sticking with low-paid, low-responsibility, or nondegree positions for more than a couple of years can put you behind professionally.

Finally, develop and respect your own ideas and follow your passions and dreams. Ignore those who say your ideals are impractical, but work pragmatically to develop your dream. Passionate and interested people find their way into interesting, satisfying careers. The main things they have in common are that they work hard and don't give up. Always pursue dreams, never chase money. If you focus always on the work you want to do, the money to do it will find you.

RESOURCES

American Fisheries Society, www.fisheries.org

Blue Ocean Institute, www.blueoceaninstitute.org

Fish Base, www.fishbase.org

Marine Conservation Biology Institute, www.mcbi.org

Marine Stewardship Council, www.msc.org

National Coalition for Marine Conservation, www.savethefish.org

National Marine Fisheries Service, www.nmfs.noaa.gov

Pew Fellows Program in Marine Conservation, www.pewmarine.org

SeaWeb, www.seaweb.org

U.S. Fish and Wildlife Service National Fisheries Program,
   http://northeast.fws.gov/fisheries

World Wildlife Fund Endangered Seas Program, www.panda.org/endangeredseas

13. FORESTRY

Dr. Jerry Franklin. Source: Mary
Levin, University of Washington.

A CONVERSATION WITH
Jerry Franklin

Dr. Jerry Franklin's self-proclaimed mission is to "cut the best deal
I can for forests and trees in a world that's dominated by
humans." Franklin knew by the time he was nine years old that
he wanted to be a forester, and he has since become one of the
most respected experts in the field, often referred to by the
media as "the guru of old growth." He is credited with modify-
ing tree-harvesting practices so that foresters can retain, rather
than "cut," biodiversity.

Today, Franklin can often be found tramping and working in the
forests he loves. He is currently professor of ecosystem analysis at
the College of Forest Resources at the University of Washington
and a fellow of the American Association for the Advancement
of Science. A prominent scientist and forest advocate, Franklin
has written over four hundred articles for scientific journals and
international, federal, and state agencies. He is the co-author,
with David Lindenmayer, of *Towards Forest Sustainability* (Island
Press, 2003).

What Is the Issue?

The earth has an astounding range of forest ecosystems covering four billion acres of land. These forests vary wildly by topography, weather, and many other factors. In addition, we find that the pressures on forests in wealthy nations are starkly different from those in developing nations.

When portraying forestry today, it is rather easy to conjure up the image of an unstoppable juggernaut of greedy loggers systematically destroying forests and leaving behind only smoldering stumps and shopping malls. The reality, though, is much more complex. It's also more optimistic. It's undisputable to me, for example, that North American forests and their ecosystem functions are generally healthy, thanks to improvements in land use practices, scientific understanding, and technology. It is important that we see realities clearly and speak about them honestly.

These caveats aside, it is possible to identify a short list of concerns facing the world's forests, including:

1. Deforestation
2. Forest fragmentation, degradation, and biological simplification
3. Air pollution
4. Pests, pathogens, and invasive species
5. Fire and fuels management

There are many more problems facing the world's forests (or made worse by declines in forest health), but these five are the keys to understanding the threats to our forests, as well as the work ahead of us to protect them.

• Deforestation

Deforestation is not a single process. We might distinguish between three different types:

• Loss of original virgin forests
• Decline in ecologically intact and relatively undisturbed "frontier" forests
• Variations in total forest cover, including new growth from reforestation, plantations, and gradual regrowth of forested areas.

The World Resources Institute estimates we have lost almost half of the global forests of 8,000 years ago and that less than half of what remains could be called intact "frontier" forests. Forty-eight percent of these remaining treasures are boreal forests that lie in a broad belt of primarily coniferous trees between the Arctic tundra and the temperate zone. Forty-four percent are tropical forests. Brazil, Canada, and Russia contain nearly 70 percent of these forests. Deforestation is still very much with us. Although the rate may have slowed recently, observers estimate that the world loses anywhere from ten to sixteen million hectares of the forest every year.

No matter how we define deforestation, it's an overwhelming problem in developing nations, especially those in tropical regions. Studies indicate that expansion of subsistence farming is perhaps the leading cause of forest loss in Africa and parts of Asia, and that converting forestland to large-scale ranching and agriculture is a major cause in Latin America and selected Asian nations. Building roads is a critical factor that often accelerates deforestation. Trade in tropical woods, however—sometimes mentioned as a driver of deforestation— is mostly confined to markets in Japan, Korea, China, and Taiwan. In fact, demand for wood products generally is not a major cause of deforestation.

• Forest Fragmentation, Degradation, and Simplification

When forests are extensive, intact, and relatively undisturbed, they provide invaluable "ecological services." Healthy forests act as sinks to absorb tons of excess carbon dioxide, potentially reducing global warming. They protect soil and purify the air and water by regulating the hydrologic cycle, retaining essential nutrients, and assuring proper sediment production and distribution—all of which keeps ecosystems in balance. They are essential to the protection of native biodiversity, including so-called charismatic mammals, which are usually large and often roam across vast areas in search of food. Forests provide habitat for at least half, and perhaps as many as 90 percent, of the planet's terrestrial species.

When forests are fragmented, degraded, or simplified, these ecological services are diminished—although the extent of this impact is hotly debated. It's certainly true, for instance, that a roadless forest wilderness is ecologically preferable, but we may be able to manage the scale, placement, and design of forest roads to minimize impact. One thing is clear—even degraded forests provide critically important watershed protection and wildlife habitat. One-third of the United States, for instance, is forested, even though we have lost most of our old growth.

Forest plantations are both part of the problem and part of the solution. There is no question that plantations simplify natural ecological systems because they lack the biodiversity of old forests. It's also clear, however, that plantations' exceptionally high productivity allows us to leave other forests alone, while still meeting rising demand for wood products.

- Air Pollution

Atmospheric pollutants directly affect forest health—especially in the United States and Europe. Acid rain, caused largely by emissions from power plants in the eastern United States, has had devastating impacts on entire forest ecosystems. High levels of ozone also cause serious forest problems in some regions.

Global climate change is a unique air pollution issue. Forest ecosystems have adapted over long periods of time to specific locations with unique seasonal variations in rainfall, temperature, and other climactic factors. If climate changes rapidly, forests that were well-attuned to particular regions may be unable to adapt, with possible die-offs due to heat, drought, decreased resistance, or competing species. Some forests, on the other hand, may flourish and expand with climate change. Global climate change could also trigger large-scale, catastrophic disasters, such as increased intense wildfires and windstorms.

- Invasive Species

The introduction of pests, pathogens, and organisms from other continents has played a major role in the degradation of forests in the United States—in fact, there is probably no greater threat to forests on this continent. Exotic pests have killed eastern and western white pines, Fraser fir, eastern hemlock, and flowering dogwood in many parts of their ranges. As many as 400 exotic insects and approximately twenty-four known disease-causing pathogens are doing real harm. Almost all important commercial tree species are now subject to pests and pathogens, and the rate at which these organisms are being introduced is accelerating. The Asian long-horned beetle provides an example. It's a serious threat to maple, beech, and birch forests on up to 48 million acres from New England to the Midwest. It also endangers poplar and aspen forests.

Pests and pathogens come from many sources. Shipments of green plants and raw wood between continents, for instance, bring exotics that can wreak havoc, causing scientists to call for prohibitions to slow these destructive hitchhikers. Products made from wood that has not been kiln-dried and treated—wooden shipping crates, for example—are also serious problems.

- Fire and Fuels Management

In 2000, widespread wildfires burned across 8.4 million acres in the American West. In 2002, 6.7 million acres burned. These were the largest fire seasons in the past fifty years. One culprit implicated in these costly disasters was the unnatural accumulation of fuels that cause fires in western pine and mixed conifer forests, which put these areas at uncharacteristic but severe risk. The buildup of forest fuels is the result of fire suppression, past logging practices, and the creation of dense tree plantations. Correcting these fuel accumulations presents an exceptional management challenge.

How Are Environmental Professionals Approaching the Issue?

Nothing can bring back our lost virgin forests, but if society adopts forestry practices built on principles like the six components listed in Box 13-1 and links them with creative economic development strategies in poor nations, we can protect and enhance what remains and usher in an age of ecological forest practice.

- Building a Better Understanding of Forest Ecology

Before professionals can provide effective forest stewardship, we must first understand forests as complex ecological systems that include human beings.

BOX 13-1

STRATEGIES FOR SUSTAINABLE FORESTRY

- Building a Better Understanding of Forest Ecology
- Developing Better Policy for Ecosystem Management
- Implementing Ecological Forest Management Practices
- Making "Fiber Farms" Work
- Creating Eco-Certification Standards
- Forging Public-Private Partnerships

This means a thorough grounding in environmental science and a rapid further evolution of that science. We must literally see the forest for the trees.

We've been on this path for a short amount of time. In fact, the core concepts now driving changes in forest stewardship practices would not have been possible without our increased understanding of forest structures and how they have co-evolved with wildlife habitats, ecosystem functions, and human activities. Forest science is a field in which better theoretical understanding leads directly to better practical management. Both ecological researchers and **foresters** owe this improvement to the development of sophisticated Geographic Information Systems, Global Positioning Systems, remote sensing data from satellites, and other monitoring and analytical tools.

· Developing Better Policy for Ecosystem Management

Even if the principles of environmental science were not driving us to manage forests ecologically, government policy and financial considerations would probably require us to do so. Forests are subject to regulation and international protection standards aimed at endangered species, air quality, watershed protection, wilderness preservation, wildfire risk management, and much more. State governments also provide regulatory guidance for forestry activities on lands within their jurisdiction.

However, the result of these cumulative responsibilities has been *de facto* demand for ecosystem management of forests. Although the federal government itself has rapidly developed a strong capacity to manage its properties with sensitivity to ecological values, state regulation has been much less likely to reflect the best knowledge of forest function, and laws governing private land are even less demanding. At its best, this forest policy mix is contradictory, confusing, and expensive to administer. Laws like the Endangered Species Act were developed to address individual problems and then layered onto an already complex legal system. Therefore, a more integrated forest policy regime based explicitly on ecosystem management principles is a strategic necessity.

When creating new policies, policy makers should use economic incentives to stimulate the adoption of ecologically beneficial management practices. For example, targeted tax incentives could encourage stewardship, particularly on private land—as is done when farmers are paid for soil conservation. Government could also offer incentives to those who protect fish habitats and retain stands of trees needed by endangered species. Government could also spur

FORESTER

AT A GLANCE

Employment:

35,000 (including related range and soil scientists)

Demand:

Slow

Breakdown:

Public sector, 40 percent

Private sector, 60 percent

Trends:

• A wave of government forestry employees will retire between 2004 and 2010, creating many job openings.

• Increased demand for sustainable forestry will create jobs in the private timber industry for people trained in the skills needed to achieve certification.

Salary:

Entry-level salaries begin around $32,000, with annual median earnings reaching $52,000. Most employees earn between $40,145 and $64,440 with top-level foresters averaging $72,000.

JOB DESCRIPTION

Forests cover nearly 750 million acres in the United States. Foresters manage and oversee these lands, ensuring that woodlands remain in abundance even as they supply commercial and industrial production. A forester's greatest challenge is determining how to comply with environmental regulations and maintain forest health while meeting economic and public consumption needs.

Foresters protect public forests, but also procure timber for manufactured products such as wood and paper. "These needs are sometimes at odds with each other, as many foresters now practice sustainable forestry," says Michelle Harvey, director of science and education at the Society for American Foresters. "This work involves balancing conservation and procurement by considering the economic and environmental impacts of forestry on natural resources and ecosystems."

Foresters meet regularly with landowners, loggers, government officials, and interest groups to determine objectives, negotiate timber sales, manage contracts, develop management plans, and meet environmental regulations. They also plant and grow trees. During the regeneration process, foresters sometimes find signs of disease or insects on trees, and then work to protect healthy trees from infestation.

A physically demanding profession, forestry requires that professionals spend most of their working time outdoors. This requires physical fitness—enough to climb up steep slopes and walk long distances to perform the work. Foresters use sophisticated instruments like clinometers, diameter tapes, increment borers, and bark gauges to measure tree growth. They also utilize remote sensing, photogrammetry, and Geographic Information Systems to make maps, evaluate forest growth, and simulate the effects of clear-cutting and forest fires. Most foresters work for private industry or for state and federal government agencies, while others serve as independent consultants.

GETTING STARTED

• A bachelor's degree in forestry is required for employment; programs accredited by the Society of American Foresters (www.safnet.org) are highly regarded.

• A master's or doctorate degree is also recommended for advanced positions in forest entomology, urban forestry, and forest resource management.

• Sixteen states either require licensing or encourage voluntary registration for foresters who wish to earn the title of "professional forester" and practice forestry.

• Certification opportunities are available, such as the Society of American Foresters' Certified Forester® program.

ADVICE FROM THE PROS

Jan Davis, a heritage and legacy forester for the Texas Forest Service in College Station, Texas, says this field offers work in forest protection, research, education, urban preservation, and much more. "You need to remain flexible and open-minded about the opportunities out there," she says. "Look at all the possibilities and get active in civic organizations that relate to your interests in forestry. One thing's for sure: foresters in all backgrounds have a lot of personal satisfaction from their work, knowing that they're preserving a vital resource in our country."

investment in markets for forest services such as carbon sequestration and watershed protection.

Creating a scientifically based national carbon policy is one specific example of how we could promote sustainable forestry while simultaneously mitigating impacts caused by cutting forests. Unfortunately, drafters of the Kyoto Protocol did not choose to award carbon benefits for leaving forests in place (including old-growth forests), nor for moving to longer rotations that would provide much greater carbon storage in routinely harvested forests. These are positive incentives, however, and we would be wise to adopt them.

- ## Implementing Ecological Forest Management Practices

Managing forests for ecosystem health means adopting new practices on the ground. Simply moving away from practices like clear-cutting and toward variable retention harvesting systems—as many public and private forest managers have been required to do—has helped immensely. This type of harvesting leaves behind organisms and healthier soil structures that help regenerate damaged ecosystems more quickly. Similarly, selection management processes, where **foresters** remove only some trees so that overall forest cover is maintained, can allow us to harvest timber, yet preserve high levels of ecological function and habitat.

Ecosystem-based approaches offer strategic solutions. Western forests, for example, must protect endangered salmon, preserve hydrological values, enhance wildlife habitat, and protect against wildfires at the same time. One linchpin to all of these objectives turns out to be the maintenance of healthy watercourses, critical elements of which are streamside protection and restoration. By investing in these elements, we can meet a range of goals.

One of the promising arenas for ecological forestry practices is forest restoration, which is both a strategy for forest health and a growing professional specialty. Restoration specialists who work to bring forests back from destructive practices or devastating fires provide opportunities to measure the effects of our practices against clear baselines, which enhances learning. Many new forest scientists are drawn to restoration work because it offers a chance to make significant contributions simultaneously to basic science and conservation.

Additionally, in our ongoing battle against tropical deforestation, a fundamental question is how to align the needs of advancing development, indigenous people, and the natural forest. Economic development strategies can help, such as ecotourism and new approaches to agriculture that minimize forest loss. Developing nations are recognizing that forests mean more than just wood. They may be able to leapfrog over the period of reckless destruction that characterized the American frontier experience and move directly to ecological forest management.

And the American experience seems to support this hope. It was not so long ago that **foresters** were overwhelmingly trained to value wood production over all other forest values. Such an attitude is not prevalent today.

Simply stating the need for ecological forestry practices, however, does not make them happen. One remaining issue is who will pay for this ecological forest management. Timber sales have been a source of forest management revenue for all public and private forest owners. As economic productivity is pushed lower into the mix of priorities (which also includes wildlife, recreation, watershed protection, and carbon absorption services), what will the source of money be? Ecological forest management does not benefit society if voters, legislatures, and consumers will not pay for it.

Making the case for ecological management requires an economic valuation of healthy forests—at least as insurance against costly disasters like floods and out-of-control wildfires. It doesn't help forest agencies if nature is only valued as "free" or "priceless." Forest managers, engineers, and ecologists are seeking the help of environmental ecologists to place monetary values on ecological health so we can estimate the cost of degraded forests and assess taxes and fees.

New sources of revenue for ecological management of U.S. forests will be needed for another reason. Wood production is moving to other countries, inevitably resulting in job losses here for skilled workers. The elimination of processing facilities and economic subsidies are also problems as the industry declines in many parts of North America. In the short term, this decline may make it more difficult to maintain desired levels of ecological services and biodiversity in our forests.

• Making "Fiber Farms" Work

Globalization of the wood products industry has accelerated the move away from traditional forestry management practices toward agronomic or agricultural approaches. Fiber farming is an extremely efficient way to produce wood, rather than relying on native forests to produce a variety of goods. Fiber farming produces large amounts of forest products at low production costs, mainly by growing trees in intensively managed forest plantations with shorter rotations. This approach is common in the southern hemisphere—where popular species include radiata or Monterey Pine, and some of Australia's eucalyptus trees—and it is expanding rapidly. Between 1980 and 1995, forest plantations in the developed world increased from 50 million to over 90 million hectares, and growth has continued. In the developing world, plantation acreage doubled to 81 million hectares in the same period. More than 80 percent of plantations in the developing world are in Asia.

Many people view fiber farming as a positive strategy because it does not entail harvesting natural forest ecosystems for wood. One ecological benefit is that most fiber farms have been established on lands that are not currently occupied by forests, such as abandoned grazing and agricultural lands in the temperate and subtropical regions of Australia, New Zealand, Chile, Argentina, southeastern Brazil, and South Africa. These regions usually have developed economies and necessary technology for managing wood plantations. Hence, fiber farming need not impact tropical forests or harm indigenous populations.

- Creating Eco-Certification Standards

At the end of 2003, the Forest Stewardship Council (FSC) announced that Domtar, a major Canadian company, had agreed to certify all 22 million acres of its forests and mills to the FSC's high sustainability standards. This decision more than doubled the total amount of North American forestland that is "eco-certified" by the FSC and raised the total amount of certified forestland to 25 percent of Canada's commercial forests. The Domtar announcement comes on the heels of high profile decisions by Home Depot, Lowe's, and Ikea—the world's largest purchasers of wood—to require eco-certification of the wood sold in stores.

Certification systems like FSC and, to a lesser degree, the industry-sponsored Sustainable Forestry Initiative give consumers assurance that the wood they buy is grown, harvested, and manufactured in socially responsible ways that do not create environmental degradation. When we demand eco-certified wood, it increases market demand for sustainable forestry, resulting in increased acreages coming under ecological forestry management practices.

Certification systems like the FSC's serve another important purpose. As

REAL ESTATE ASSOCIATE

AT A GLANCE

Employment:

Approximately 18,000 to 20,000 jobs in 2003

Demand:

Falling

Breakdown:

Public sector, 75 percent

Nonprofit, 25 percent

Trends:

• Job employment in the public sector is decreasing, as federal and state agencies struggle with the economy's downturn.

• Reduced government funding affects professionals' ability to acquire land for restoration and redevelopment purposes.

Salary:

Entry-level positions earn from $25,000 to $30,000, with median salaries reaching $60,000 a year. Top salaries can be $90,000 or more a year.

JOB DESCRIPTION

Real estate associates secure lands for public land trusts and brownfields redevelopment firms to preserve as open space or clean up and turn into a viable developments. Each year, they help preserve more than 500,000 acres from development, according to the Land Trust Alliance (www.lta.org). These are "highly motivated individuals who have the skills to make opportunities and negotiate," says Jack Steffurd, assistant director of Stewardship and Protection at Natural Lands Trust. "So much of land protection lies in identifying the need for conservation and then pursuing that need to its end."

Professionals in this field can work for organizations such as the U.S. Fish & Wildlife Service or the National Parks Service (NPS), often holding the title of "land protection specialist" or "realty specialist." They draw on strong negotiating and grant writing skills and often are self-starters who can pursue projects on their own. Their main objective is to acquire either the land or its development rights for preservation. Other realty specialists focus their efforts on acquiring brownfields sites with the intent of restoring and reselling them for development or preserving the land as open space.

In their work, real estate associates first identify landowners with property that fits well in the organization's open space portfolio. This involves studying title deeds and interviewing property owners. They must also ensure that the land up for consideration meets federal and state environmental laws. The next step is negotiating with landowners interested in donating or selling their property for conservation purposes. Real estate associates often call on attorneys to settle transaction details with landowners.

To formally acquire lands, professionals must also secure private or public funding to cover the cost of buying land or creating conservation easements. To accomplish this, they apply to local, county, state, and federal agencies for grant funding.

GETTING STARTED

• A bachelor's degree is the minimum requirement for employment, but a master's degree is highly recommended for those interested in managerial positions.
• Look for programs in natural resources management, conservation, or land use planning.
• While it is not necessary for employment, some training in negotiating real estate transactions can be helpful.
• Most positions require some experience. Look for positions that will introduce you to grant writing and negotiation.

ADVICE FROM THE PROS

David Funk, a realty specialist for the NPS's Philadelphia office, says his job is to acquire privately owned lands, which the NPS then preserves as parks or wildlife refuges. Funk suggests that prospective professionals focus on the meticulous details of this work. "You must be careful how you use words and craft legal documents, so good knowledge of laws and solid writing skills are a must. A bit of training in real estate or law can't hurt, either."

Career Spotlight

The Gifford Pinchot National Forest
Vancouver, Washington

The U.S. Department of Agriculture Forest Service is a big landowner and a big employer. Taken together, the service's 155 national forests and twenty grasslands make up 8.5 percent of the total land area in the United States, or 191 million acres—an area equivalent to the entire state of Texas. Managing the national forest system involves over 30,000 people in forty-four states, making the Forest Service the nation's largest single employer of conservation professionals.

One of the very first forests protected by the agency in the early 1900s was the 1.4-million-acre Gifford Pinchot National Forest in southwest Washington State. Rising majestically within the forest is the snow-capped and blown apart peak of Mount St. Helens—centerpiece of the sprawling 110,000-acre National Volcanic Monument.

Many people associate the national forest system with logging, and it's true that plenty of people are employed as foresters and forestry technicians who do just that. It can come as a surprise, however, to learn how much of the workforce is devoted to preserving national forests for outdoor recreation and wildlife habitat, instead of timber.

For Diane Bedell, time in the forest is literally a day at the office. Bedell is an outdoor recreation planner and wilderness manager on a Pinchot ranger district—one of 600 such districts in the national system. Among other duties, she's responsible for four wilderness areas that are part of the National Wilderness Preservation System.

"Managing wilderness is a balancing act," she muses. "We have to preserve wilderness character while providing for an appropriate level of recreation opportunities. I define the wilderness program goals and develop programs to reach them. This means understanding what motivates our visitors and staying current with constantly evolving outdoor recreation trends."

Bedell's education includes a B.A. from Eastern Washington University and an M.A. from the University of Minnesota, both in parks and recreation administration. Bedell says a minor in business administration was an important addition. And she practiced what she preaches. "I spent five years leading adventure trips and teaching backpacking, rock climbing, canoeing, whitewater rafting, and dog sledding."

That combination of serious academic training and an interest in the natural world is typical of Gifford Pinchot employees—like Mario Isaias-Vera. "I love the outdoors," Isaias-Vera says, "and so I enjoy being an engineer here." Born in Mexico, he immigrated to the United States in 1991 with a B.S. in chemical engineering and since has earned two master's degrees—in environmental education from Southern Oregon University and environmental engineering from Oregon State. He was hired by the Forest Service as an environmental engineer in the Pinchot forest and now provides guidance for managing hazardous materials programs, coordinating cleanups and emergency responses, designing civil engineering projects, and complying with environmental regulations.

Almost all actions in Gifford Pinchot are carefully regulated and none more so than

Mario Isaias-Vera in the field.

activities that could adversely affect the Pacific salmon, whose habitats are nearly all protected under the federal Endangered Species Act. Protecting the forest's fish is the work of Diana Perez. Perez started out as a summer intern for the service as an undergraduate and came back to a forest in Oregon every year while completing a B.S. in wildlife management from Texas Tech University and an M.S. in fisheries science from Oregon State University. She transferred to Pinchot in 2000 and now works as a forest fish program manager, performing biological evaluations of forest management actions to determine effects on fish habitat and developing plans to avoid negative impacts.

Forest Service managers use the word "partnership" a lot, and Perez is passionate about their necessity. "Coordinating with the state Fish and Wildlife Department and Trout Unlimited is an integral part of doing what's best for the fish. This work costs quite a bit of money, and partnerships are vital to restoration."

Like Bedell, Perez notes that "people skills" are important complements to scientific training —and essential for everyone at the Forest Service. "Program management, budget development, and knowledge of Forest Service policy are all immensely valuable." she says. "But outstanding interpersonal skills are absolutely critical, because gaining public support is ultimately the key to success."

consumers demand wood that meets FSC criteria, forestry practices may form around the new standard—providing a model and increasing the urgency for others to act. The very existence of the standard often pushes even noncertified companies toward better practices. For example, as was the case with organic food, competing wood certification systems will hopefully lead to a call for an "official" eco-certification standard from government. Debate over certification standards will be an important part of sustainable forestry efforts.

• Forging Public-Private Partnerships

The public no longer wants, or allows, forest management to be given over to forest managers and engineers. Even if it did, federal and state laws now require open processes and citizen involvement for nearly all significant decisions, and activists consistently press for more.

Effective **foresters** have embraced public involvement as a major part of the solution to our current forest issues. Forestry involves the social sciences at least as much as the ecological sciences. **Foresters** and conservation specialists have, by necessity, become savvy managers of people and politics as well as trees.

Perhaps the best example of public involvement is the routine use of public-private partnerships in forest management. Forestry professionals simply cannot manage forest issues alone. Leaders in the field are building alliances and coalitions that involve conservationists, hunters and fishermen, government agencies, community groups, land trusts, schoolchildren, and universities. Forest supporters even share daily management work with the public. In many areas volunteers have planted native species in reforestation efforts, repaired damaged lands after fires, built and maintained hiking trails, and conducted biological inventories. Organizing these successful collaborations—and bringing the community into the forest—is one of the happiest tasks of forest management.

Career Advice from the Expert

Today, **foresters** and forest technicians working on innovative and protective management practices must be able to combine knowledge with practical management skills. There's a large demand for people who can function as planners and knowledge integrators, because forestry work increasingly involves multi-disciplinary stakeholders. Understanding and managing living forests now requires the ability to analyze myriad factors that interact with one another across varying aspects of time and space. Once we begin thinking about forests as watershed protection systems, habitats for wildlife, parks for hikers, hedges against global warming, and tree farms that won't be harvested until years from now, we will fundamentally change the desired qualifications of forest professionals. For this reason, prospective employees should be able to use Geographic Information Systems (GIS), integrate and synthesize information, lead planning efforts, and work collaboratively with many groups.

A bachelor's degree in a natural or applied science is the minimum requirement for this field. A master's degree in forestry or related applied science will give prospective forestry employees a real advantage. Additionally, experience in science or research areas related to ecosystem management will provide a sound basis for future employment.

Aspiring forestry professionals should look for ecological forestry work in privately held companies, trusts, and nongovernmental organizations that manage forestlands, as well as local, state, and federal government agencies that have forest stewardship responsibilities. The forest community is large and diverse—new professionals would do well to look beyond the "name brand" employers for opportunities.

RESOURCES

American Forest and Paper Association, www.afandpa.org

Forest Stewardship Council, www.fscoax.org

Gifford Pinchot National Forest, http://www.fs.fed.us/gpnf/

Global Forest Watch, www.globalforestwatch.org

Society of American Foresters, www.safnet.org

United Nations Forum on Forests, www.un.org/esa/sustdev/forests.htm

United States Department of Agriculture Forest Service, www.fs.gov

World Bank/ World Wildlife Fund Alliance for Forest Conservation and Sustainable Use, www.forest-alliance.org

World Resources Institute Forest Frontier Initiative, www.wri.org/wri/ffi

14. FRESHWATER QUALITY, USE, AND PROTECTION

Sandra Postel on a lake in western Maine.

A CONVERSATION WITH
Sandra Postel

For twenty years, Sandra Postel has tracked global water trends
and their implications for society. "The degree of water scarcity
we now see around the world is staggering," she says, noting
that satisfying the world's thirst while "protecting the health of
the aquatic environment that sustains all terrestrial life" is one of
our greatest challenges.

Postel is director of the Global Water Policy Project in Amherst,
Massachusetts, and a visiting senior lecturer at Mount Holyoke
College. In 2002, she was named one of the "Scientific American
50" by *Scientific American* magazine.

A prolific writer, Postel is author of *Pillar of Sand: Can the Irrigation Miracle Last?* (W.W. Norton, 1999). Her article "Troubled
Waters" was selected for the 2001 edition of *Best American Science and Nature Writing*. Most recently she co-authored, with
Brian Richter, *Rivers for Life* (Island Press, 2003) and, with Amy
Vickers, "Boosting Water Productivity" in *State of the World
2004* (W.W. Norton and Worldwatch Institute, 2004).

What Is the Issue?

The earth's freshwater supply is finite, but human demands for freshwater have grown—and continue to grow—steadily. In 2015, nearly 3 billion people—40 percent of the projected world population—will live in countries that find it difficult or impossible to mobilize enough freshwater to satisfy the food, industrial, and domestic needs of their people. As a result, competition for water will increase—between cities and farms, between states and provinces, and between nations that share a common river. At the same time, more water devoted to human purposes will leave less for the natural world itself, eventually resulting in the collapse of ecosystems.

• Increasing Human Demands

With the rise in population and economic growth since 1950, we saw a parallel surge in human demands for water, energy, irrigation, and flood control. We met these demands by building more and larger water projects, especially dams, river diversions, and groundwater wells. Since 1950, the number of large dams (those at least fifteen meters high) has climbed from 5,000 to 45,000—an average of two new large dams a day for the last half century.

• Devastating Environmental Impacts

These structures have dramatically altered the natural river flows and habitats that support freshwater life. Scientists estimate that at least 20 percent of earth's 10 thousand freshwater fish species are at risk of extinction or are already extinct. In the United States, The Nature Conservancy estimates that 69 percent of freshwater mussels are (to some degree) at risk of extinction, as are 37 percent of freshwater fish species and 36 percent of amphibian species. We are losing valuable goods and services that healthy freshwater ecosystems provide for us including fish to eat, pure water to drink, recreational enjoyment, and many other valuable benefits.

At the same time, we have to wonder where additional water will come from to meet new human demands. Many major rivers—including the Colorado, Ganges, Indus, Rio Grande, and Yellow—now run dry for portions of the year. Water tables are falling from overpumping groundwater in large portions of China, India, Iran, Mexico, the Middle East, North Africa, Saudi Arabia, and the United States. We might think of this as a kind of hydrologic

Table 14-1. Life-Support Services Provided by Rivers, Wetlands and Other Freshwater Ecosystems

Ecosystem Service	Benefits
Provision of water supplies	• More than 99% of irrigation, industrial, and household water supplies worldwide come from natural freshwater systems
Provision of food	• Fish, waterfowl, mussels, clams, etc. are important food sources for people and wildlife
Water purification/waste treatment	• Wetlands filter and break down pollutants, protecting water quality
Flood mitigation	• Healthy watersheds and floodplains absorb rainwater and river flows, reducing flood damage
Drought mitigation	• Healthy watersheds, floodplains, and wetlands absorb rainwater, slow runoff, and help recharge groundwater
Provision of habitat	• Rivers, streams, floodplains, and wetlands provide homes and breeding sites for fish, birds, wildlife, and numerous other species
Soil fertility maintenance	• Healthy river-floodplain systems constantly renew the fertility of surrounding soils
Nutrient delivery	• Rivers carry nutrient-rich sediment to deltas and estuaries, helping maintain their productivity
Maintenance of coastal salinity zones	• Freshwater flows maintain the salinity gradients of deltas and coastal marine environments, a key to their biological richness and productivity
Provision of beauty and life-fulfilling values	• Natural rivers and waterscapes are sources of inspiration and deep cultural and spiritual values; their beauty enhances the quality of human life
Recreational opportunities	• Swimming, fishing, hunting, boating, wildlife viewing, water-side hiking and picnicking
Biodiversity conservation	• Diverse assemblages of species perform the work of nature (including all the services in this table), upon which societies depend; conserving genetic diversity preserves options for the future

Source: S. Postel and B. Richter, *Rivers for Life*, Island Press, 2003.

deficit financing—borrowing tomorrow's water to meet today's needs. But what happens when tomorrow comes?

Freshwater has no substitute in most of its uses. We can think about replacing our reliance on coal and oil with renewable energy sources, like solar and wind power. But not so for water. It is essential for growing crops, for manufacturing material goods, and for drinking, cooking, and other household functions.

• Dimensions of the Problem

In a nutshell, the challenge we face is to determine how to meet human needs for water sustainably while simultaneously protecting the freshwater ecosystems that support all life—including ours. This is a big challenge. It may help to break the problem down into how we use—and abuse—water.

Agricultural Use

Agriculture accounts for about 70 percent of all water removed from rivers, lakes, and underground aquifers worldwide. In much of the world, growing crops requires irrigation—either year-round or, at minimum, during the dry season. Irrigation is critical to global food production: only 18 percent of the world's cropland is irrigated, but that irrigated land produces 40 percent of the world's food. A large portion of irrigation water is stored behind dams and diverted through canals, but never benefits crops. Some of it seeps through unlined canals, runs off farmers' fields, or simply evaporates before reaching crop roots. Some of that "lost" water may return to a stream or recharge groundwater and get used by another farmer. The inefficiency is costly, however. River flows are altered excessively, aquatic habitats are destroyed unnecessarily, and more freshwater is contaminated by fertilizers and pesticides.

Many poor farmers have no irrigation options at all. Sadly and ironically, most of the world's 800 million hungry people live or work on farms. Without irrigation water, they simply cannot grow enough food to support their families. Yet most of the conventional irrigation technologies are too expensive for them to purchase.

Industrial Use

Industries account for about 22 percent of global water use, but a much higher share (nearly 60 percent) in industrial countries. Virtually every product we buy—from paper and clothes to cars and computers—requires water to manufacture. Industries also generate large volumes of wastewater, and (especially in developing countries) this wastewater is often released untreated into nearby rivers and streams, contaminating scarce water supplies.

Household Use

Cities and towns account for less than 10 percent of world water use. Because this use is very concentrated geographically, though, it places great pressure on local water bodies. Treating and distributing drinking water is also expensive, as is collecting and treating urban wastewater.

In the wealthy United States, the average resident uses too much water—100 gallons a day, much of it to water thirsty backyard lawns in climates where those lawns do not belong. In poor countries, a total of 1.1 billion people don't have enough water—they lack access to a safe supply of drinking water, which results in widespread disease and as many as three million preventable deaths each year. At the same time, in both rich and poor countries, many urban water systems lose 30 to 50 percent (or more) of their water supplies to leakage.

Pollution

Pollutants from all three sectors, including fertilizers, pesticides, heavy metals, and organic chemicals, contaminate rivers, lakes, and aquifers. A study of 139 U.S. streams by the U.S. Geological Survey found that 80 percent of them had traces of at least one drug, endocrine-disrupting hormone, insecticide, or other chemical—some at levels known to harm fish and other aquatic life. For more information about harmful effects of chemicals, see chapter 19 about toxics and human health.

How Are Environmental Professionals Approaching the Issue?

There are three overarching pieces to the global freshwater challenge: maintaining ecosystem health, boosting water productivity and efficiency, and extending water access to the poor. Each of these efforts involves work by scientists, resource managers, economists, engineers, planners, and policy strategists. Box 14-1 gives some of the ways environmental professionals are tackling the challenge.

• Allocating Water for Ecosystem Health

Hydrologists now know that healthy rivers require not just a minimum quantity and quality of water, but a pattern of water flow that resembles their natural flow regime. This means establishing what scientists call an "environmental flow prescription"—a tool that enables resource managers to be sure that a river system gets water in the right amounts and at the right times to safeguard its health. This is a complicated task. River ecologists have made great strides over the last decade in developing methods to determine how much water a river needs.

There is also, of course, a policy component to this task. New laws and regulations may be required to ensure that freshwater ecosystems get the water they need. South Africa's 1998 water law, for example, calls for meeting the basic water requirements of both people and ecosystems as the highest priority for water allocation. Officials in Australia have capped water extractions from the Murray-Darling basin to secure more water for the river system.

BOX 14-1

STRATEGIES FOR PRESERVING OUR FRESHWATER SOURCES AND MEETING HUMAN NEEDS

- Allocating Water for Ecosystem Health
- Restoring Rivers
- Boosting Agricultural Water Efficiency
- Bringing Irrigation Technologies to Poor Farmers
- Meeting Water and Sanitation Targets
- Implementing Urban Conservation Programs
- Adopting Regulations and Economic Incentives
- Encouraging Water Diplomacy

- Restoring Rivers

Around the world a burgeoning movement of river restoration activities is gathering steam. According to The Nature Conservancy, some degree of flow restoration is already taking place in more than 350 rivers around the world. Dams are being taken down, levees are being set back to reconnect rivers with their floodplains, and dam operations are being modified to re-create predam river flow patterns and critical habitats.

The good news coming from these restoration efforts is that, when given a chance, rivers do heal. Just a year after the removal of Edwards Dam on Maine's Kennebec River, anadromous fish like Atlantic salmon and sturgeon, striped bass, and alewifes returned to the river in great abundance. In October 2003, negotiations among a partnership formed by the PPL Corporation, a private hydroelectric power producer; conservation organizations, including Penobscot Nation, American Rivers, the Atlantic Salmon Federation, and Natural Resources Council of Maine; and government agencies including the U.S. Fish and Wildlife Service, the Maine Departments of Inland Fisheries and Wildlife and Marine Resources, and the Maine Atlantic Salmon Commission led to a path-breaking decision to tear down two dams and bypass another to give wild Atlantic salmon a comeback chance in Maine's Penobscot River. Civil and **environmental engineers** and

WATER RESOURCES MANAGER

AT A GLANCE

Employment:

90,000

Demand:

Good

Breakdown:

Public sector, 80 percent

Private sector, 20 percent

Trends:

• Population and economic growth will increase demand for additional water and wastewater treatment plants.

• Employment in the private sector will grow faster than average, as industry deregulation is increasing reliance on private firms that specialize in managing and operating water facilities.

Salary:

Starting salaries begin around $29,000 and median earnings are $46,000. The highest salaries are around $64,000 for nonmanagers and much higher for top managers.

JOB DESCRIPTION

"Simply put, some water resources managers make water safe to drink, and others return purified wastewater back into rivers, streams, and the ocean," says Dr. Christopher L. Lant, executive director of the Universities Council on Water Resources (www.uwin.siu.edu/ucowr) in Carbondale, Illinois. "Professionals must constantly test water for contaminants and ensure it meets all requirements as mandated by the Safe Drinking Water Act," adds Lant.

Water resources managers working in water treatment plants supervise teams of workers who capture water from wells and rivers. They treat that water and send it to the homes and residences of their customers. Those working in wastewater plan, design, and operate plants that purify wastewater. Both types of water resources managers monitor water use in their communities. "Some even develop plans to protect watershed health, conduct cost-benefit studies of watershed improvement projects, or draw up master plans to manage and rehabilitate watersheds," says Jonathan Yeo of the Massachusetts Water Resources Authority.

Water resources managers also negotiate with neighboring communities for water rights and with neighboring water facilities to meet their community's water supply demands. They also meet with public interest or community groups to present water resource proposals. Accordingly, water resources managers need strong analytical and communication skills. They also rely heavily on computer programs such as Geographic Information Systems (GIS) or Global Positioning Systems (GPS) to compile data and develop plans. Most water resources managers work in the public works departments of state or local governments, while a smaller number work for private water supply and sanitary service companies.

GETTING STARTED

• A bachelor's degree in civil or environmental engineering, chemistry, hydrology, or environmental studies is recommended; postsecondary training is increasingly becoming an asset as treatment plants grow more complex.

• Should you decide to work at a water or wastewater treatment plant after high school, some employers pay partial tuition for related college courses in science or engineering.

• Forty-nine states require certification exams. Many state water department agencies also offer training to improve applicants' skills and knowledge for the job.

• Learn computer skills and become familiar with Geographic Information Systems, as water resources managers must work with more computer-controlled equipment than ever before.

• Government agencies, such as the U.S. Geological Survey (www.usgs.gov) and local water departments, often offer internship opportunities and career information.

ADVICE FROM THE PROS

Thomas Newton, water department supervisor for the city of Easthampton, Massachusetts, says he earned his first job for the water department, because "I was in the right place at the right time. For most, it's important to start working for a local water department to get out and see what it involves. The hard part is getting your foot in the door. Being persistent, patient, and certified will be an advantage."

hydrologists are essential for completing such technical tasks after **lawyers**, activists, and **policy analysts** hammer out hard-won agreements. Conservation scientists now expect the salmon to rebound rapidly to ten times their present depleted number in the river.

• Boosting Agricultural Water Efficiency

Because agriculture accounts for the lion's share of global water consumption, using irrigation water more productively is crucial to meeting our global water challenges. The various ways to do this are shown in Table 14-2. One technology gaining ground is drip irrigation, which delivers water directly to the roots of plants at low volumes. The practice typically saves water and improves harvests at the same time—often doubling or tripling water productivity.

About half of the world's people prefer rice as their staple food, and 90 percent of it is grown in Asia, where many rivers and aquifers are overtapped. Researchers and farmers working in the fields, however, have found that the traditional practice of keeping rice fields flooded throughout the season is not essential for high yields. By applying a thinner layer of water or letting the field dry out between irrigations, farmers can in some cases reduce water applications by 40 to 70 percent without significantly reducing their yields.

Innovative programs that match real-time weather conditions to crop water requirements are also making a difference. Remote weather stations feed temperature, humidity, wind speed, and other data to a centralized computer. Farmers can tap into the system to determine how much water they should apply to their fields, depending upon the crop they are growing and the moisture already stored in the soil. Information technology programs of this sort operate in parts of California, Texas, and the Pacific Northwest.

• Bringing Irrigation Technologies to Poor Farmers

Some exciting efforts are under way to bring the benefits of irrigation to poor farmers. The keys to success are affordable technologies designed for small plots of land, combined with the development of a manufacturing and marketing network to supply the technology. One model of success is the treadle pump, a human-powered device that lifts shallow groundwater to the surface. Farmers in Bangladesh have purchased about 1.2 million of them. For a cost of $35, each pump increases farmer income an average of $100 in the first year. Farmers I spoke with in Bangladesh said they were not only no longer hungry, they were now taking crops to market.

ENVIRONMENTAL ENGINEER

AT A GLANCE

Employment:

52,000

Demand:

Strong

Breakdown:

Public sector, 30 percent

Private sector, 50 percent

Academia, 20 percent

Trends:

• While there are more jobs than environmental engineers to fill them, the work often changes with changes in government policy and public concerns.

• Surveys show that the number of environmental and civil engineering graduates are declining, creating more of a demand for these professionals as industry continues to face increasingly complex government regulations.

Salary:

Entry-level engineers with B.S. degrees earn up to $51,000, while engineers with a master's degree earn $58,000 or higher. Most earn between $49,000 and $80,000, with top salaries reaching $94,000 or more.

JOB DESCRIPTION

Clean drinking water, solid waste disposal, pollution control, and public health management are the handiwork of environmental engineers. "How important is it to drink water? How important is it that garbage doesn't pile up on streets? How important is it to breathe clean air?" asks William Anderson, president of the Academy of Environmental Engineers (www.aaee.net) in describing their work's importance.

Environmental engineers differ from other kinds of engineers in that they research and analyze environmental problems and concerns, and then design systems to address those issues. They commonly design wastewater and hazardous waste treatment plants, develop and support systems that treat drinking water, and monitor air quality in local communities.

Some engineers may travel far and wide, as many developing countries need experts to implement infrastructures to meet drinking water, wastewater treatment, and air quality needs. Other engineers tackle projects in industry, redefining the way companies manufacture products by redesigning plants and production processes that use sustainable or environmentally sound materials.

Tim Groninger, an environmental engineer who specializes in water systems, is an engineer who works in industry. He uses computer and mathematical modeling systems, programming languages such as Fortran and Visual Basic, Geographic Information Systems, and spreadsheet programs to redesign industrial plants.

In addition, environmental engineers also conduct environmental impact assessments and write technical reports for their clients. They often work with other engineers, outside clients, architects, contractors, and, at times, community members to implement systems that address problems and adhere to federal and state regulations. They work primarily at private consulting and research firms, government agencies, and major corporations. Others may work in academic settings, teaching or conducting research. Most environmental engineers have academic backgrounds in civil, environmental, geological, mechanical, chemical, and biological engineering.

GETTING STARTED

• A bachelor's degree is the minimum requirement for employment, though many employers prefer engineers with master's degrees and even doctorates that focus on environmental, chemical, or civil engineering. For a list of programs accredited by the Accreditation Board for Engineering and Technology, visit the board's website (www.abet.org).
• The National Society of Professional Engineers (www.nspe.org) offers information about professional engineering licensure, which significantly increases salary earnings.
• Look for internships or job opportunities at your local public works department or a private engineering firm.

ADVICE FROM THE PROS

Dan Hatfield is an environmental engineer for the URS Corporation in Omaha, Nebraska, which assists the U.S. Army Corps of Engineers in cleaning up polluted military sites across the country. "I suggest getting a civil engineering degree to fall back on," he says. "Environmental engineering changes with the political climate and the tides of public opinion, and isn't always stable, whereas people always need roads and buildings."

Table 14-2. Menu of Options for Improving Irrigation Water Productivity

Category	Option or Measure
Technical	• Land leveling to apply water more uniformly
	• Surge irrigation to improve water distribution
	• Efficient sprinklers to apply water more uniformly
	• Low-energy precision application sprinklers to cut evaporation and wind drift losses
	• Furrow diking to promote soil infiltration and reduce runoff
	• Drip irrigation to cut evaporation and other water losses and to increase crop yields
Managerial	• Better irrigation scheduling
	• Improving canal operations for timely deliveries
	• Applying water when most crucial to a crop's yield
	• Water-conserving tillage and field preparation methods
	• Better maintenance of canals and equipment
	• Recycling drainage and tail water
Institutional	• Establishing water user organizations for better involvement of farmers and collection of fees
	• Reducing irrigation subsidies and/or introducing conservation-oriented pricing
	• Establishing legal framework for efficient and equitable water markets
	• Fostering rural infrastructure for private-sector dissemination of efficient technologies
	• Better training and extension efforts
Agronomic	• Selecting crop varieties with high yields per liter of transpired water
	• Intercropping to maximize use of soil moisture
	• Better matching crops to climate conditions and the quality of water available
	• Sequencing crops to maximize output under conditions of soil and water salinity
	• Selecting drought-tolerant crops where water is scarce or unreliable
	• Breeding water-efficient crop varieties

International Development Enterprises, a Colorado-based organization with offices around the world, is expanding its experience in Bangladesh into a global effort called the "Smallholder Irrigation Market Initiative." Development specialists working on this project aim to lift millions of rural farm families out of poverty through access to affordable irrigation.

- Meeting Water and Sanitation Targets

Periodically governments, donor agencies, development organizations, and nongovernmental organizations come together and rededicate themselves to the critical challenge of meeting the safe drinking water and sanitation needs of the poor. In 2000, the United Nations General Assembly adopted as one of its Millennium Development Goals to reduce the proportion of people lacking access to safe drinking water by half. At the World Summit on Sustainable Development in Johannesburg, South Africa, in 2002, nations committed to cutting the proportion of people lacking adequate sanitation by half. Stepped-up funding and technical support for water and sanitation projects should expand opportunities for NGOs, government agencies, international aid organizations, and community groups working to meet these targets.

- Implementing Urban Conservation Programs

Conserving water used to be something we thought of doing only during a drought. However, as water supplies tighten, saving water in our cities and homes is a necessity—and it often saves energy, money, and habitat at the same time.

Most successful urban conservation programs involve a mix of measures—including fixing leaks in the distribution network itself, installing efficient fixtures and appliances in homes and offices, auditing industries to see how they can reduce or reuse their water, and educating people about ways they can save water. Water use efficiency analysts have helped the Massachusetts Water Resources Authority (MWRA), which supplies more than 2 million people in metropolitan Boston, reduce water demands on its system by about 25 percent—allowing it to cancel a controversial plan to build another dam on the Connecticut River and saving its customers more than $500 million in capital expenditures. Like many of the nation's several hundred utilities, the MWRA calls on a core staff of specially trained and certified water and wastewater engineers and technicians, supplemented by a wide array of management, information technology, and legal talent to achieve goals like this.

In the United States, a new frontier in conservation is "water-thrifty" landscaping. Our love affair with green lawns results in using nearly 8 billion gallons a day to irrigate turf and grass. But memberships in natural landscaping organizations such as the American Society for Landscape Architects are growing, and **landscape architects** are working with a number of corporations to switch to attractive native landscapes at some facilities. Besides saving

ENVIRONMENTAL REGULATOR

AT A GLANCE

Employment:

66,000 (including some environmental scientists)

Demand:

Falling

Breakdown:

Public sector, 100 percent

Trends:

- Although state government environmental protection agencies employ over half of all environmental regulators, they are now facing devastating budget shortfalls, reducing employment opportunities.
- Because regulators often act as educators and environmental law enforcers, employers seek professionals who can both teach people in a tactful manner and strictly enforce regulations.

Salary:

Entry-level positions are $31,000, with median salaries reaching $50,000. Top salaries exceed $81,000, but most earn between $38,000 and $65,000.

JOB DESCRIPTION

The Environmental Protection Agency monitors a broad range of federal regulations that limit the amount of hazardous pollution entering the air, land, and water. Environmental regulators enforce these laws to protect the health of plants, animals, aquatic life, the environment, and people. Regulators can work for federal agencies or at the state level, usually in environmental protection or ecology departments.

As part of their job, environmental regulators grant permits to industrial and municipal facilities to ensure that they do not exceed preset levels of pollution discharge. Regulators can require a company or municipality to buy environmental friendly equipment for their facilities if they exceed recommended emission levels. They may also play a part in alerting the media and the public to threatening problems such as overly high levels of air pollution, toxic spills, or poor water quality issues.

Overall, regulation is highly technical work, which often involves collecting samples and analyzing data. It requires familiarity with scientific sampling as well as compiling and calculating statistics. For example, at the Wyoming Department of Environmental Quality, environmental regulator Leah Krafft reviews permit applications, researches environmental history, evaluates water quality data, and compares pollution levels to federal and state regulations.

Paul Hadley, a regulatory agent for the California Department of Toxic Substances Control in Sacramento, focuses his efforts on groundwater and soil cleanup processes. "Not everything works at first," he says, "but it's critical work, so you just have to try again until you get it. It plays a part in taking care of our environment."

No matter where they work, environmental regulators often collaborate with other state and federal environmental protection offices, engineers, landowners, and private industry to ensure that everyone is doing their part to control pollution levels.

GETTING STARTED

• A bachelor's degree in a field such as biology, chemistry, hydrology, or geology is the minimum education requirement, though many environmental regulators have master's degrees or doctorates.
• Strong computer skills are needed to work with sampling data, document research, and extensive Internet searches.
• Relevant career information can be found at the Environmental Protection Agency's website (www.epa.gov).

ADVICE FROM THE PROS

Don Kerstetter, chief of continuance compliance in air quality for Pennsylvania's Department of Environmental Protection, spends his days regulating the air emissions of hospitals and manufacturers with incinerators. "We have to make sure these groups are following the law," he says. Kerstetter advises that newcomers be aware that this work involves perseverance and commitment, since it involves constant attention and monitoring. "You have to keep after folks and not get discouraged. Persistence is one thing you must have to work in this field."

Career Spotlight

Lower Colorado River Authority
Austin, Texas

The Colorado River in Texas—not to be confused with the one that carved the Grand Canyon—flows through Austin from the western part of the state on its way to the Gulf of Mexico. Along the way, the Colorado provides electric power to more than a million people, quenches the thirst of growing cities, sustains wildlife and fish habitats, waters crops and livestock, and offers recreation for the thousands of people annually who enjoy the popular Highland Lakes chain of six reservoirs.

Six hundred miles of the river is protected and managed by the 2,200 employees of the Lower Colorado River Authority (LCRA)—a workforce that includes professional engineers, hydrologists, biologists, park managers, and recreation specialists. Formed in 1934 to reduce flooding and bring economic development, LCRA today owns and manages parks and nature preserves, water and wastewater systems, recreational facilities, a series of dams, and electric power plants.

Maintaining a healthy ecosystem is job one at LCRA, and a major focus is on keeping pollutants from runoff out of the river. Rachel Andrews, a nonpoint-source inspector for the agency, explains: "Seventy percent of water pollution can be attributed to 'nonpoint' sources, and it's far cheaper to prevent pollu-

tion than to remediate a damaged ecosystem. My job is to minimize the impact to water quality that comes from development activities." Andrews earned a bachelor's degree in civil engineering and Spanish from the University of Texas at Austin and notes that both are essential in her work.

Andrews helps ensure that best management practices (BMPs) are implemented and maintained to reduce pollutant load in the Highland Lakes. Examples of BMPs include rock berms and silt fences used during construction, and detention ponds and vegetative buffers for postconstruction.

"One of the new techniques we are using to minimize nonpoint-source pollution is porous pavement," she says. "We also encourage the use of a 'sheet flow and filter' system so that water runoff is filtered through grass instead of channeling down curbs."

Senior aquatic scientist Jerry Guajardo also helps protect the waters of the lower Colorado River. Coming to LCRA with a bachelor's degree in biology from Southwest Texas State University (now Texas State University), he ensures that the organization has quality water data and that municipalities and industries do not cause damage to the river ecosystem.

"The importance of water can't be overstated," Guajardo says. "In Texas, water is more valuable than gold, and with the increasing population, there is a growing emphasis on having good, clean water for use. People now are starting to ask more questions about where their tap water comes from and about the quality of that water source. We need this data to answer their questions."

Nora Mullarkey, senior water conservation coordinator, helps the public use valuable drinking water wisely. Mullarkey earned her master's degree in public health from the University of Texas Health Sciences Center at Houston and previously worked for the city of Austin, performing water conservation evaluations and running a citywide low-flow showerhead retrofit program.

"How water is used reflects an area's culture," Mullarkey says. "We try to understand that culture and explain to people how water conservation is connected to environmental protection. Using water wisely, for example, saves energy used to store, save, and provide drinking water."

At LCRA's McKinney Roughs Nature Park near Bastrop, schoolchildren participate in some of the state's most innovative educational programs. The idea is to educate kids about the environment while letting them have a good time canoeing, swimming, and enjoying the river—and Cody Ackermann is the man who brings Texas children and the Colorado together. A recreation program director for LCRA's nature parks and preserves, Ackermann received a bachelor's degree in environmental and resource management from Southwest Texas State University (now Texas State University) before coming to LCRA.

"LCRA's nature parks provide a great opportunity for children who don't often get the chance to enjoy the outdoors," Ackermann says. "The kids are enthusiastic about learning more of how nature interacts with their everyday lives, and it's rewarding knowing they are taking knowledge back with them that can help perpetuate a new generation of environmentally aware people."

LCRA scientists take regular water quality measurements on the Lower Colorado River to guarantee clean, healthy water for people and wildlife.

water, native landscaping reduces the use of pesticides and the health risks they pose.

- Adopting Regulations and Economic Incentives

It's great to have efficient technologies, good scientific tools, and an arsenal of conservation measures—but will they get used? That's where laws, regulations, and incentives come into play. Without a requirement that dam operators release flows from reservoirs in ways that benefit the river ecosystem, it probably won't happen. Increased use of drip irrigation would not likely occur without reductions in subsidies for irrigation water. Also, pricing water in ways that discourage excessive use—for instance, hefty summer surcharges to discourage heavy lawn watering—can promote conservation.

Regulations have critical roles to play too. Limits on groundwater extractions are urgently needed in much of the world so that aquifers don't get over-pumped. U.S. cities and towns are now benefiting from federal water efficiency standards that came into effect in 1994, requiring all new toilets, faucets, and showerheads to meet specific flow or flush rates. Water conservation expert Amy Vickers estimates that by 2020, these efficiency standards will be saving a volume of water equal to that used by four to six cities the size of New York City.

- Encouraging Water Diplomacy

Preventive diplomacy is an increasingly important facet of equitable and sustainable water use. As competition over water intensifies within and between countries, it's critical to bring feuding parties to the negotiating table before tensions escalate. These conflicts may involve farmers versus environmentalists over whether to allocate scarce water to fish or farming, dam operators versus conservationists over how to harmonize use of a river for economic purposes with new ecological goals, or one country versus another over how to equitably share a common river. River basin commissions and other institutions that bring parties together to work out their differences can solve our more contentious water problems through cooperation.

Career Advice from the Expert

I sometimes describe my work as examining the world through a water lens. Water problems are inherently interdisciplinary, which means the door of employment opportunities is open to people trained in many different fields and with many different interests. The challenge is to match one's skills and passions with an opportunity to improve the world's water situation.

Practical experience can help prepare people for careers in the water field. Many developing countries face unique challenges of providing safe drinking water to rural villagers and irrigation water to small farmers. Spending time working in the field—for instance, with a Peace Corps program—can provide very valuable experience with these challenging goals. Such opportunities may also expand technical skills in setting up irrigation systems and building small rainwater harvesting systems, as well as developing educational and training skills.

Internships and entry-level jobs that provide opportunities to work closely with more seasoned professionals in the field can be enormously helpful. My first "real" job was at Natural Resources, a small consulting firm where I learned about the nuts and bolts of water conservation planning and policy by working with a senior engineer. Organizations like American Rivers and the National Fish and Wildlife Foundation offer internships that give students and recent graduates the opportunity to work with freshwater professionals on projects such as floodplain protection, policy reforms on Capitol Hill, and conservation programs. Time spent working for state and federal government agencies will introduce prospective professionals to how government systems function and policy decisions are made.

Last, I would urge anyone interested in the water field to hone their writing, speaking, and communication skills. Even the best idea is unlikely to fly if one doesn't pitch it effectively to the right audience.

RESOURCES

American Rivers, www.americanrivers.org

American Water Works Association, www.awwa.org

Hydrology and Water Resources Programme of the World Meteorological Organization, www.wmo.ch/web/homs/1stpage.html

International Development Enterprises, www.ide-international.org

International Hydrological Programme of the United Nations Educational Scientific Cultural Organization, www.unesco.org/water/ihp/index.shtml

International Rivers Network, www.irn.org

International Water Management Institute, www.cgiar.org/iwmi

Lower Colorado River Authority, http://www.lcra.org

The Nature Conservancy's Freshwater Initiative, www.freshwaters.org

U.S. Geological Survey, www.usgs.gov

World Commission on Dams, www.dams.org

15. GREENING OF BUSINESS

Dr. Stuart Hart.

A CONVERSATION WITH
Stuart Hart

As a pioneer in the field of business and sustainability, Stuart Hart is all too familiar with companies' traditional views of the environment as raw material to exploit for profit, or a lawsuit waiting to happen. The challenge of global sustainability, however, is changing forever how companies operate.

"Businesses now have the opportunity and challenge to achieve sustainability instead of degradation and exploitation," Hart asserts. "More and more companies realize they can reduce pollution, solve social problems, and increase profits simultaneously—and a small handful of the best corporations are doing just that." Throughout his career as an academic, consultant, and management educator, Hart has helped leading corporations accomplish these goals.

Hart earned his M.S. in environmental management from Yale University and his Ph.D. in strategy from the University of Michigan. He is currently professor of strategic management and co-director of the Center for Sustainable Enterprise at the University of North Carolina's Kenan-Flagler Business School. He also recently accepted the S.C. Johnson Chair in Sustainable Global Enterprise at Cornell's Johnson School of Management. His influence is wide-ranging in academia and business, as evidenced by more than fifty published works, including "Beyond Greening: Strategies for a Sustainable World," which won the 1997 McKinsey Award for Best Article in the *Harvard Business Review*.

What Is the Issue?

Most businesses still view social and environmental issues largely as added costs and afterthoughts. Companies construct and execute their competitive strategies and only then deal with whatever environmental or social problems occur as a result, usually only to the degree that ensures their ability to maintain legitimacy and their right to operate. While today's leading-edge companies are working to improve products and operating processes as well as innovate clean technologies, most focus only on incremental improvements or cleanup activities after the impact or damage has occurred. This approach is neither sustainable nor profitable.

• Environmental Degradation

Mounting biodiversity loss, climate change, increased waste, and suburban sprawl can all be traced to corporate production and consumer activities. The earth is being stressed beyond its carrying capacity. We see this in our depleted farmlands, fisheries, and forests; choking air pollution; poverty; disease; and unsustainable population growth. All of these serious problems can be traced in some way or another to exploitative or degrading corporate activities.

Moreover, American corporations have helped to build a consumer culture that often encourages excessive consumption. Americans by far have the largest ecological footprint, a measurement that defines the natural resources used in our production and consumption patterns. Through globalization, we are increasingly exporting these consumption patterns to other parts of the world, meaning that more people are adopting excessive and wasteful habits. If everyone on the earth consumed like Americans, the planet would, in all likelihood, be pushed beyond its ability to support life as we know it.

• Failure of Emerging Market Strategy

Over the past ten years, multinational corporations (MNCs) have taken existing wasteful products and production systems, which previously served well-off people in developed countries like the United States, Japan, and those in Western Europe, and extended them to the elites and rising middle classes in the developing world. MNCs tweak their business systems only incrementally to do this, using access to raw materials and cheap labor to lower production costs. Not surprisingly, these strategies have spawned increases in environmental degradation throughout the developing world.

Interestingly, these emerging market strategies have not been particularly successful. Elites and rising middle classes in the developing world do not have the same tastes as Americans, Western Europeans, and Japanese. In addition, it has been more difficult than anticipated to transplant technology, products, and business systems to the developing world. As a result, many large companies have backed away from their emerging marketing strategies. This opens the door to a more creative (and potentially sustainable) way of approaching these markets.

- ## The Inequality Gap

While the globalization of corporate investment has produced widespread benefits in certain countries or regions—like East Asia, for example, where some people's quality of life has improved significantly—it has not distributed benefits equally to most people in the world. Rather, the rich have become richer, while quality of life for the poor has not improved at a similar rate or has become even worse, leading to growing inequity between those at the top of the global economic pyramid and those at the bottom, who are excluded or exploited in the process. Figure 15-1 shows which people comprise the tiers of the global economic pyramid.

MNC operating practices that rely on resource extraction and increased environmental degradation often significantly affect the "survival economy," those people who live directly off nature in rural villages at the bottom of the economic pyramid. Four billion people, or two-thirds of the global population, live in the survival economy. They are often illiterate, have no formal education, participate in the informal economy, and are under increasing pressure to migrate to urban areas because of environmental degradation, population pressure, or lack of opportunity.

Unfortunately, urban migration is not usually a successful solution. Few people who search for wage jobs in factories actually find work. Instead, most remain in the informal economy in urban slums and shantytowns. This presents

Annual Per Capita Income*	Tiers	Population in Millions
More Than $20,000	1	75–100
$1,500–$20,000	2 & 3	1,500–1,750
Less Than $1,500	4	4,000

* Based on purchasing power parity in U.S.$
Source: U.N. World Development Reports

FIGURE 15-1. The World Economic Pyramid

an interesting challenge for corporations. On the one hand, markets at the top of the pyramid are stagnant. On the other, the majority of people in the world have unmet needs or are being actively exploited by vendors that charge exorbitant prices or provide poor-quality products and services. In addition, there is growing hostility toward large companies that are primarily exploiting resources and cheap labor to serve the rich. This further exacerbates the growing gap between the wealthy and the desperately impoverished, who have few options for survival other than directly living off the land, often in environmentally degrading ways.

How Are Environmental Professionals Approaching the Issue?

There is an enormous opportunity for large companies to serve those people who have been excluded from or negatively impacted by economic globalization. And corporations can improve their business practices in the developed world to be less damaging to the environment and use fewer natural resources. Pollution prevention, product stewardship, changes in technology, credit, cost, and distribution, as well as a corporate sustainability vision, are critical for accomplishing these goals. Large corporations with global reach have the technological, managerial, and financial resources to dip into the well of innovations needed to profit from these opportunities. Consequently, business is the critical sector that can bring about sustainable change.

There appear to be two types of companies that are pursuing proactive strategies. The first group includes corporations that are improving current products and operating systems by incorporating social and environmental issues into product and process design. They think in terms of waste avoidance and pollution prevention, for example. They operate more efficiently and pro-

---

BOX 15-1

GREENING OF BUSINESS STRATEGIES

- Preventing Pollution
- Taking a Product Stewardship Approach
- Developing Clean Technology
- Inventing a Sustainability Vision

duce less waste of all types, which lowers cost and reduces risk. They also include a broader array of stakeholders in their decision making, which ensures legitimacy and reputation, and improves the environmental performance of product systems. Most leading-edge companies today are part of this group and are making significant headway in bringing environmental and social issues into mainstream business operations.

The other group represents the future of business. A very small number of companies are in this category. They do not stop at making incremental improvements to current products, processes, or product systems. Rather, they are geared more toward innovation and acquiring entirely new skill sets, technologies, and capabilities that will allow them to leapfrog into the future. New capabilities may disrupt or even creatively destroy current products and product systems. By employing these capabilities, for example, companies will ultimately enter new markets that importantly serve the bottom of the global economic pyramid and have huge potential to improve quality of life for all people. The strategies associated with these two groups are discussed in more detail below.

- ## Preventing Pollution

Pollution prevention means changing internal processes to avoid the production of waste, rather than cleaning it up at the "end of the pipe" after it's already been created. By employing this strategy, companies can lower costs, increase business performance, and simultaneously increase efficiency and profitability. Pollution prevention is by far the most widely distributed "greening of business" strategy, as most major companies in the world today utilize it, typically through environment, health, and safety departments. Indeed, those who do not put themselves at a serious disadvantage. However, by simply implementing pollution controls, companies spend more money on unnecessary and unproductive technology, rather than using pollution prevention to improve their overall processes. 3M is widely recognized as one the pioneers in pollution prevention. Since the 1970s, the company has saved more than $700 million through its "Pollution Prevention Pays" program, by eliminating pollution at the source through product reformulation, process modification, equipment redesign, and recycling and reuse of waste materials. Since then, the concept of "eco-efficiency" has become the main platform for industrial groups like the World Business Council for Sustainable Development (WBCSD), a coalition of 170 international companies that provides business leadership as a catalyst for change toward sustainable development. WBCSD also promotes the role of eco-efficiency, innovation, and corporate social responsibility.

305

ENVIRONMENT, HEALTH, AND SAFETY MANAGER

AT A GLANCE

Employment:

36,500

Demand:

Slow

Breakdown:

Private sector, 80 percent

Public sector, 10 percent

Academia, 10 percent

Trends:

• Increasing demand exists for environment, health, and safety managers as federal and state governments require companies to meet and exceed environmental and public safety regulations.

• Entry-level positions are abundant, but middle- and upper-level positions are hard to come by.

Salary:

Entry-level salaries begin at $39,000, with median annual earnings from $49,500 to $73,000. Top-level positions earn $90,000 or more.

JOB DESCRIPTION

While industry produces the goods and services that people demand, it can also create serious environmental hazards in the process. Environment, health, and safety (EHS) managers develop programs and processes to ensure that companies comply with federal, state, and local environmental laws, or go beyond them to achieve high levels of environmental excellence.

"We focus on regulatory compliance, ensuring the companies we oversee keep current with certification as well as create products that are safe for the environment," says Paul L. Dadak, environment, health, and safety program manager for Risk Management and Workplace Services.

To accomplish these goals, EHS managers must know the details of environmental regulations, including the Clean Air and Clean Water Acts, the Pollution Prevention Act, and the Toxic Substances Control Act. They often work with environmental regulators and other EHS managers to determine the extent of contamination and the effects of exposure on environmental and public health through air, water, and soil sampling.

George Larsen, former EHS manager at Lockheed Martin Aeronauticals, says that most EHS managers have scientific and engineering backgrounds. They often draw upon this experience to determine the causes of problems and offer long-term solutions—for example, when addressing the effects of toxins on environmental and public health.

Because industries are profit-driven, EHS managers must communicate the importance and business value of environmental and public safety to their company's upper-level management. They often do so by producing financial reports that prove decreased incidence of illness or injury, as well as positive environmental track records for impacting companies' bottom lines. EHS managers are also catalysts for reducing the number of regulated chemicals used within specific industries, which makes production processes safer for employees, customers, and people who live close to industrial facilities.

GETTING STARTED

• A bachelor's degree in environmental engineering, industrial hygiene, safety engineering, or environmental science is the minimum requirement for employment. Managerial positions often demand a master's degree.
• Working on air, water, and soil sampling in laboratories or regulatory agencies offers practical experience for this occupation.
• Computer proficiency and writing skills are necessary for preparing technical and financial reports.

ADVICE FROM THE PROS

Ann Reiter, senior environmental, health, and safety specialist for Biogen, Inc., says her job involves limiting the company's scientists from producing too many products. "I find I have to hone my communication skills because you have to talk to people at different levels who don't always want to listen," she says. "It's also important to understand how finances work, because when you assign a monetary value to a proposal, you get people's attention."

- ## Taking a Product Stewardship Approach

Product stewardship focuses on improving the environmental or social performance of the entire product system. Here, researchers, designers, and technical experts look at products and systems on a life cycle basis and then develop new designs and technologies to make appropriate changes to create more sustainable business models. Marketing and product developers then infuse life cycle management, development, and design processes with marketing strategies in order to offer the best products and realize the full benefits of this strategy.

Collins and Aikman (C&A), now part of the Tandus group, is a carpet manufacturer that has demonstrated product stewardship by redesigning commercial carpet, normally comprised of vinyl backing and nylon fibers. Over the past ten years, C&A has designed a proprietary system to take back existing PVC-backed commercial carpet in order to remanufacture new carpet backing. The company worked with DuPont to dramatically reduce the amount of nylon needed for face fibers and uses this reduced content in its remanufactured product. Additionally, the remanufactured product has more stability, a tighter weave, and added cushion, making it more functional than using virgin PVC versions of the same product and improving its environmental performance.

C&A's success in taking back old carpet, incorporating it into new backing, and changing the product design dramatically has demonstrated the profitability of product stewardship. The company has successfully reduced cost and improved sales, growing 12 to 15 percent annually over the last decade in an industry that is growing at only 3 to 4 percent per year.

- ## Developing Clean Technology

As economist Joseph Schumpeter would describe it, clean technology seeks to creatively destroy what currently exists through fundamental redesign of products and product systems, rather than just continuously improving existing technologies. For this reason, clean technology involves entirely new approaches that can better serve human needs. To do this, businesses must determine for whom and what purpose they are developing products.

Early markets for clean technology will probably not be mainstream, established markets. The United States, for example, has dragged its feet in implementing renewable energy technology for decades, although the technology has been around equally as long. This is not surprising. The typical American is already extremely well served by cheap energy from nonrenewable and centralized sources.

INDUSTRIAL ECOLOGIST

AT A GLANCE

Employment:

1,200

Demand:

Rising

Breakdown:

Academia, 70 percent

Private sector, 15 percent

Public sector, 10 percent

Nonprofits, 5 percent

Trends:

• As this new field evolves there are increasing numbers of academic positions offered at colleges and universities that focus on industrial ecology studies.

• Employers look for professionals with a background in engineering, the natural sciences, or some related field, so they can apply such knowledge to industrial ecology studies.

Salary:

Starting salaries range from $35,000 to $50,000, and top wages exceed $110,000.

JOB DESCRIPTION

While industries like manufacturing and production help stabilize the economy, they also can cause environmental problems such as air pollution and hazardous waste buildup. Industrial ecologists are a diverse group of researchers, chemists, engineers, educators, outreach coordinators, lawyers, and others looking to identify ways that industry can still maintain production while interacting safely with nature. They accomplish this by using natural ecosystems as models for innovative industrial systems, explains Beverly Chevalier, program coordinator for the International Society of Industrial Ecology at Yale University's School of Forestry and Environmental Studies. "Industrial ecologists base their work on a holistic, systems view, requiring the participation and expertise of many disciplines to develop innovative solutions to complicated environmental problems," she says.

As most industries extract and process raw materials and fossil fuels, eventually dumping waste back into natural systems, industrial ecologists seek to redesign such processes and conserve natural resources. Simultaneously, industrial ecologists try to improve a company's resource productivity and competitiveness, and lower production costs. They incorporate technological innovation and consumer consumption patterns into their designs by using material and energy flow studies, life cycle assessment, and organizational design models. Industrial ecologists also strive to bring environmental concerns to the attention of elected officials and policy makers.

These professionals draw from many different academic fields for their work, including natural resource management, ecology, engineering, economics, public policy, and public health. They can work for private consulting firms, colleges and universities, corporations, nonprofit organizations, private and public foundations, and government agencies. Some of their job titles include environmental engineer, natural resources lawyer, college professor, research associate, project manager, and environmental planner.

GETTING STARTED

- A bachelor's degree in environmental engineering, natural resource management, or economics is an excellent starting point for aspiring industrial ecologists.
- A master's degree or Ph.D. is highly recommended, as most industrial ecologists perform advanced research or work in academia. The International Society for Industrial Ecology (www.yale.edu/is4ie) lists programs that offer study in this field.
- Reading publications like the *Journal of Industrial Ecology* (www.mitpress.mit.edu) will keep you abreast of current trends.

ADVICE FROM THE PROS

Kristan Cockerill, adjunct lecturer for the University of New Mexico's American Studies department, says there are increasing numbers of positions being offered in academia, specifically called or connected to the principles of industrial ecology. "There are few degrees that are explicitly for industrial ecology, however, even though some employers expect you to have industrial ecology-relevant skills. One approach is to major in whatever is your passion and then pursue industrial ecology-relevant research."

We have no motivation to adopt renewable or distributed energy sources such as solar, fuel cells, or micro turbines. In fact, our current infrastructure (including an interconnected grid of fossil-fuel-burning power plants) works against their adoption. This will be the case until clean technologies are clearly superior—in that they function better and are available at cheaper cost—to current technology.

In contrast, companies might have greater success in introducing such technologies at the bottom of the pyramid. Using clean technologies in undeveloped markets will allow people living without basic services and infrastructure to leapfrog the developed world's inefficiencies. For example, because no electricity or energy infrastructure exists at the bottom of the pyramid, the survival economy can immediately utilize clean technologies—distributed generation and renewable energy sources such as solar, for example—without much resistance.

Specifically, light-emitting diode (LED) technology might enable the bottom of the pyramid to leapfrog lightbulbs. This technology is 90 percent more energy efficient than incandescent bulbs and lasts for eight to ten years, all of which will give communities in the developing world considerable advantages. This technology can also be affordable. By combining white LED arrays with a small solar photovoltaic collector, battery, and wiring, a rural lighting system could cost as little as thirty dollars.

Given the apparent failure of aid-based models and the International Monetary Fund's (IMF) structural adjustment policy, which imposes austerity measures to ensure countries repay loans and restructure their economies, a commerce-led approach to sustainable development might be the best option for the future. Combining clean technology with creative business models that involve collaboration with local people, nongovernmental organizations (NGO), and small-scale entrepreneurs will allow businesses to create solutions to social and environmental problems that will, for the first time, lift the base of the pyramid to generate income and improve quality of life.

With a little imagination, for example, companies might create a micro-finance platform for products like LEDs, suddenly creating 2.5 billion customers. Creating local businesses will allow larger companies to improve efficiency and create cheaper products and production capacities because they will have expanding global customer bases and dramatically lower overheads.

- Inventing a Sustainability Vision

Rather than continue to use existing business approaches that successfully serve the top of the pyramid, businesses must creatively invent a new sustainability

GLANCE

Employment:

Over 100,000 jobs in the environmental and alternative energy industry

Demand:

Rising

Breakdown:

Private sector, 100 percent

Trends:

- Entrepreneurs' success is affected by public opinion and economic progress.
- More entrepreneurs will find profitable business niches as demand for environmentally friendly products increases.

Salary:

$0–$1,000,000. You make only what you put into the work!

JOB DESCRIPTION

Limited only by their imagination, entrepreneurs introduce environmentally sound products and services to the market at an astounding rate. Some entrepreneurs create, package, and sell these products, while others develop products, systems, and technology with an ecological purpose. "I went into the solar energy business because of my passion for the earth and my search to find a livelihood that would support both my family and the nexus where my passions and gifts met," says Geoff Greenfield, owner of Third Sun Renewable Energy Systems, an Ohio firm that designs and installs solar energy systems. As the company "president, designer, contractor, accountant, and anything else that pops up," Greenfield says his work is stressful and overwhelming, especially when he makes mistakes and has to restart projects from the beginning. However, he adds that working for himself also provides him with the ability to set his own schedule and be in charge of his business's growth.

At times, entrepreneurs work in their office, making calls, writing project proposals, and updating financial databases. Other times, they spend time in their chosen field. Greenfield is often out and about, for example, installing solar energy panels and batteries for businesses and homes. He also meets with new clients to explain the renewable energy alternatives available for their homes and businesses.

Most entrepreneurs work extremely hard and long hours. Even in the beginning, when they are generating ideas for a business or product, they must spend a great deal of time researching and fine-tuning proposals. Entrepreneurs must then write up a detailed business plan and pitch the idea to financiers. In essence, for their businesses to succeed, entrepreneurs must commit themselves to their product's success and have a keen ability to articulate ideas to investors and the buying public.

GETTING STARTED

• Entrepreneurs have diverse backgrounds, so there is no one course of study or major required. Many spend years working in the business field, learning how companies operate and succeed, before branching out on their own.

• Study or training in areas such as environmental stewardship, social impact management, and sustainability will enhance an entrepreneur's competitive edge.

• Learn basic computer skills, such as word processing, spreadsheets, and Internet use, to keep up with emerging technologies.

ADVICE FROM THE PROS

Jerry Kay, publisher of the Environmental News Network (ENN) (www.enn.com), worked as a broadcaster for 34 years before deciding to take over ENN three years ago. From the beginning, Kay has learned something new each day. "You must be incredibly flexible and willing to learn new things all of the time," he says. "You also must be persistent and tenacious in your work. Sometimes it takes me six months to develop necessary relationships."

Kay also says that entrepreneurs must be able to communicate clearly with people and be honest and direct. "Handle criticism constructively. You may have the greatest idea, but you must listen to customers to make sure it's a marketable product they want."

Career Spotlight

Seventh Generation
Burlington, Vermont

"Nature Girl." That's the job title on Janice Shade's business card, and it speaks volumes about both Shade herself and about the company behind the nation's leading brand of nontoxic and environmentally safe household products. Known more prosaically as senior marketing manager, Shade came to Seventh Generation with a B.S. in finance from Boston University, a master's in corporate strategy from the Yale School of Management, and several years of marketing experience at corporate giants like Procter & Gamble and Welch's.

"I admit that the skills and experience I gained helped get me where I am today," she says, "but it bothered me that I was promoting consumerism at the expense of the planet." Her search for work that reflected her personal values led her to an alternative medicine firm and then to Seventh Generation. She describes her current job as "identifying people who are aligned with our vision and then educating them about ways to achieve it—the most obvious of which is to use our products." Day-to-day work involves sophisticated consumer research, developing and managing advertising and public relations campaigns, maintaining voluminous amounts of information on the company website, and suggesting products that customers seem to want.

Founded by current CEO Jeffrey Hollender, the Burlington, Vermont-based Seventh Generation has quickly grown into an industry trendsetter, with twenty-five employees and products that are found in thousands of stores throughout the nation. The company's lines of paper towels, bathroom and facial tissues, napkins, paper plates, dish and laundry soaps, household cleaners, trash bags, lightbulbs, diapers, and baby wipes are all manufactured with environmental concerns in mind. Paper products, for example, are nonchlorine bleached and made from 100 percent recycled paper, and cleaning items are completely free of chemicals deemed toxic. Revenues in 2002 were up in a down economy, and annual growth over the last five years has risen steadily. Company sales are led by its popular cleaning products and, although not in the same financial ballpark as, say, Clorox, Seventh Generation is definitely doing well by doing good.

There is a strong sense of mission at Seventh Generation, and it's fueled by its customers, according to Tonianne Paquette, the company's consumer relations specialist. "I spend my days on the phone and answering e-mails," Paquette says. "People want to know about the benefits of choosing a natural or recycled product over a conventional one and exactly how they can make a difference. To educate them, I have to educate myself, and I'm constantly learning and being reinspired."

Paquette's path to her job started in the supermarket aisles. In an effort to bring recycled and nontoxic products into her house, she tried Seventh Generation offerings and found that "they worked great—I didn't get watery, itchy eyes while cleaning my home." Armed with little more than a B.F.A. in graphic design from the Massachusetts College of Art and a desire to get out of the city of Boston, which seemed unhealthy, she pulled up stakes, moved to Vermont, and landed her current job. She feels that her artistic creativity and open-mindedness are important assets to her work.

Although Seventh Generation is a different kind of company, it still must survive in a ruthless marketplace, and that means hiring people with solid business skills. As a market research analyst, Dawn Leuschner uses syndicated data and other business intelligence to benchmark the performance of Seventh Generation products against its competitors. She synthesizes piles of raw information—placing it in context for sales and marketing teams and for strategic decision making by senior executives. An experienced market analyst with a B.A. in psychology from the University of Vermont, Leuschner values the skills she gained in the social science laboratory.

Leuschner wouldn't have many purchases to track if not for a well-organized sales system throughout the country. Sales Planning and Administration manager Rob Conboy helps direct the complex process that gets products onto the shelves. Aligned with the Seventh Generation vision, Conboy recently became the first M.B.A. graduate from the University of Vermont with an emphasis in environmental sustainability. "For me, a career has to be about more than just the bottom line. It has to be about making a difference."

vision to bring costs down dramatically to serve all people in the world. We will not be able to move toward a sustainable world if companies focus only on improving the lives of the richest 800 million people. Creative strategizing and sustainability "visioning" can also help companies overcome subsidies and public policies that actively encourage businesses to choose unsustainable behaviors. These policies usually prohibit businesses from incorporating the real costs of environmental and social damage and from pursuing sustainability agendas in the developed world.

Now is an exciting time to become an entrepreneur or M.B.A. student, because sustainability visioning is coming forward in new business models. Professionals who can make cases for pursuing new markets for pollution prevention, product stewardship, and clean technology strategies will be very successful as entrepreneurs or company officers. In either position, they must be able to sell innovative ideas in order to create and then implement compelling businesses in private equity markets.

Career Advice from the Expert

The sustainability challenge offers opportunities to current M.B.A. candidates on multiple fronts, depending on their specific backgrounds and interests. For example, the M.B.A. program at the Kenan-Flagler Business School at the University of North Carolina features a concentration in sustainable enterprise. People involved in this program have a variety of backgrounds, including degrees or experience in technology, engineering, marketing, business, social science, and the humanities.

These academic backgrounds can lead students down various paths. M.B.A.s with technical backgrounds could easily find themselves in a plant working on pollution prevention, or in research and development working on product stewardship or clean technology initiatives. Aspiring business professionals with marketing experience might have opportunities for integrating environmental and social criteria into product design and development processes. Entrepreneurs will find increasing opportunities, inside or outside of large companies, to bring leapfrog technologies forward, especially in the developing world at the base of the economic pyramid. People with experience in community or development settings or with NGOs might find opportunity to include diverse voices and stakeholders within corporations to ensure they act in responsible and productive ways.

Students should look for internships or work experience in domains in which their skills may be less developed. Someone with an engineering background or experience working in a plant, or some other technical capacity might want to intern with an NGO. Similarly, someone whose background experience is working in the social sector as a teacher or nonprofit employee might want to obtain experience in a large corporation. Building alternative experiences will be valuable no matter where one is headed, as alternative experiences give employees different perspectives, making them more desirable candidates.

Prospective environmental business professionals need not sell themselves as environment, health, and safety people or sustainability specialists, because these roles will almost certainly limit job seekers' opportunities to a very tiny segment of the job market. Although work in these types of positions is necessary, people in such staff positions often find themselves on the outside looking in. More often than not, these jobs involve reporting about environmental and sustainability initiatives, rather than actually being in the game. An M.B.A. degree provides aspiring job seekers with the tools and skill sets to be in mainstream positions in the value chain—production, product development, new market development, finance, and consulting. Knowledge and skill in pursuing sustainability goals will only make such people more competitive for such positions in the future. Indeed, in today's post-Enron, post-9/11 world, large corporations must fundamentally incorporate social and environmental issues in their vision for economic globalization. M.B.A.s with skills in environment and sustainability are just the people to champion this transformation.

RESOURCES

Beyond Grey Pinstripes, www.wri.org/wri/bschools

Business for Social Responsibility, www.bsr.org

Coalition for Environmentally Responsible Economies (CERES), www.ceres.org

Harvard Business Review, www.harvardbusinessonline.hbsp.harvard.edu

International Institute for Sustainable Development, www.iisd.ca

Investor Responsibility Research Center, www.irrc.org

Kenan-Flagler School of Business Center for Sustainable Enterprise,
    www.kenanflagler.unc.edu/sei

National Association for Environmental Management, www.naem.org

Rocky Mountain Institute, www.rmi.org

Seventh Generation, www.seventhgeneration.com/

The Natural Step, www.naturalstep.org

World Business Council for Sustainable Development, www.wbcsd.ch

16. SMART GROWTH

Don Chen at his desk at Smart Growth America. Source: Smart Growth America.

A CONVERSATION WITH
Don Chen

In most American communities, sprawl dominates our landscapes, seriously impacting quality of life and degrading the natural environment. And people are beginning to take notice. "Sprawl and traffic are out of control," Don Chen says. "The vast majority wants more open space, reliable transportation choices, and revitalized neighborhoods. The evidence shows that Americans support smarter growth, and our elected officials better start paying attention."

Chen is the founding executive director of Smart Growth America, a nationwide coalition committed to promoting better alternatives to fiscally irresponsible and environmentally harmful urban sprawl.

A noteworthy advocate for sustainable communities, Chen worked as a researcher for the Surface Transportation Policy Project, World Resources Institute, and the Rocky Mountain Institute before founding Smart Growth America. He is regarded as an international expert on land use, transportation, and environmental policy. He obtained his master's degree from the Yale School of Forestry and Environmental Studies. Chen is also author of many articles and publications, including "The Science of Smart Growth" (*Scientific American*, December 2000).

What Is the Issue?

For the better part of a century, America's landscape has been shaped by a powerful combination of government subsidies, policies and regulations, industry practices, and engineering standards that encourage suburban sprawl. For many decades, most Americans regarded these policies as a good way to help ordinary folks attain their dream of home ownership and a good quality of life. But today we're finding that they are no longer helping us attain those goals and in many cases are hurting our communities.

Runaway, haphazard sprawl is so driven by these entrenched policies that locals often feel they no longer have influence over their community's future. Voters are becoming frustrated with this powerlessness, as sprawl is increasingly seen as a contributor to many things that degrade quality of life, including traffic, loss of open space, and worsening pollution. Towns and cities nationwide are struggling to keep up with the fiscal demands of growth, including new roads, sewers, schools, and other big-ticket items. And, despite its promise of fulfilling the American dream, sprawling development has failed miserably to deliver an adequate supply of affordable homes, or the variety of homes and neighborhoods that today's American home buyers and renters demand.

• Environmental Impacts

Kaid Benfield of the Natural Resources Defense Council has argued, "All of the gains we've made in the environmental movement since the 1970s could be undermined severely if we don't do something about urban sprawl and if we don't meet the challenge of better managing growth in our regions." This sweeping assertion is no exaggeration. Sprawl's most immediate impact is the consumption of undeveloped open space and farmland. In the past decade, the United States lost roughly fifty acres of prime agricultural land per hour, jeopardizing both an economic way of life and the supply of local produce, dairy products, and meats. Some development has immediate impacts on the nation's already dwindling biodiversity. In southern California, for example, sprawling development has destroyed over 90 percent of the coastal sage ecosystem. Polluted runoff from paved roads, parking lots, and other impervious surfaces is the nation's leading threat to water quality. Numerous studies document that sprawling development patterns generate more driving (both miles traveled and number of trips) than compact communities, and therefore more air pollution and greenhouse gas emissions.

- Social Impacts

Sprawl affects environmental justice, labor, and social equity. Environmental justice leader Robert Bullard, for example, contends that "sprawl ... exacerbates school crowding, heightens urban-suburban school disparities, accelerates the decline of urban infrastructure, concentrates poverty, creates spatial mismatch between urban workers and suburban job centers, heightens racial disparities, and negatively impacts public health." The AFL-CIO passed a resolution in 2001 stating that "urban sprawl strains all working families by creating overly-long commuting times, fueling air pollution responsible for skyrocketing children's asthma rates, creating a lack of affordable housing near jobs, eroding public services, and denying workers a choice about how to get to work."

In these and similar critiques, three major factors dominate. First, the well-documented "spatial mismatch" phenomenon—increasingly long distances between low-income communities and job centers—now often includes poor access to health services, schools, shopping, and other vital community needs. Second, sprawl is tied to a wide array of negative impacts, from pollution and noise to traffic fatalities and sedentary behaviors. Third, sprawl is often criticized for perpetuating social inequities through exclusionary zoning practices and discriminatory mortgage lending commonly found in suburbs.

- Fiscal Impacts

Sprawl requires constructing more roads, sewers, water mains, sewer treatment plants, utility lines, and other basic infrastructure to service fewer people than compact development. Typically, existing residents—from central cities to established suburbs—must pay for these big-ticket items through taxes, fees, utility rates, and other subsidies, because their costs are borne at a regional level. In this way, sprawl is both inequitable and fiscally irresponsible. Although additional development usually represents some additional property tax revenues, these revenues usually do not even cover the costs of the development. For example, if a city approves 500 new single-family home units, there are additional expenses for creating new infrastructure, services, and amenities, such as schools, fire and police departments, libraries, and other public services. These community resources are expensive.

- Obstacles to Creating Sustainable Urban Areas

There are three major obstacles to overcome in order to create more sustainable urban areas (see Box 16-1). First, most U.S. communities have very few alternatives

BOX 16-1

OBSTACLES TO CREATING SUSTAINABLE URBAN AREAS

- Land Use Planning Policy and Zoning
- Transportation Policy
- Lack of Affordable Housing

to sprawl because mainstream planning and zoning regulations mandate inefficient land use patterns and design standards. Such standards often require building projects to be located on large lots on discontinuous streets. They also typically require zoning that allows for only one type of building use in a particular area, creating strongly segregated residential housing, retail areas, and office buildings. As a result, someone wishing to run a simple errand must travel by car.

Our transportation policies, geared largely toward automobile dependence, are the second obstacle. Most communities are built for personal car travel, which generates excess environmental, economic, and social costs, and effectively eliminates other transportation choices such as public transit, walking, and cycling. If transportation polices provided people with more options, we'd have more cost-effective solutions for managing traffic, boosting accessibility, and improving quality of life.

The third problem is the lack of affordable housing options for families with wide-ranging incomes. A sustainable community has a diverse range of people living within it, and a lack of affordable housing prohibits this diversity from occurring. For our lowest-income families, the supply of affordable subsidized housing, rentals, and owner-occupancy opportunities is woefully inadequate in most communities. According to the National Low Income Housing Coalition, housing costs are rising faster than wages and the cost of other goods. Without decent and affordable housing choices, low-income families must seek housing in other areas, such as deteriorated urban neighborhoods with high crime and poor services, or remote sprawling places that require long commutes. Today's housing choices are also inadequate for the elderly, who predominantly wish to live where they spent most of their lives but are unable to find a variety of housing options that meets their needs and market demands. Overall, lack of housing options, in combination with the other obstacles to creating sustainable urban areas, present great challenges for implementing much-needed alternatives to sprawl.

How Are Environmental Professionals Approaching the Issue?

In the past decade, the search for better alternatives to sprawl has kicked into high gear, mainly in the form of "smart growth." The smart growth approach favors rehabilitating and reinvesting in existing communities over subsidizing new sprawl. It prioritizes the preservation of open space and farmland, and regards sound and participatory planning as an important tool for achieving more livable communities. In addition, smart growth advocates believe we need better choices in transportation, housing, and other basic necessities that growth and land development strongly influence.

Growth that is truly smart must result in desirable outcomes, among them, boosting neighborhood livability; providing better access and less traffic; fostering thriving cities, suburbs, and towns; ensuring shared benefits and social equity; lowering costs and taxes; and preserving open spaces. To achieve these outcomes, **land use planners**, **architects**, developers, **community organizers**, public health advocates, **elected officials**, and traditional conservation practitioners are committed to ten principles for implementing alternatives, shown in Box 16-2.

- Mixing Land Uses

New development works best if it includes a mix of stores, jobs, and homes, because large single-use districts make life less convenient and require more driving. In order to create more sustainable communities, public and private developers should have more incentives and opportunities to build compact, walkable, and mixed-use neighborhoods. The efforts of **land use planners** and **architects** to reform land use planning practices, zoning regulations, and design standards will lead to construction or reconstruction of more communities that maximize convenience and accessibility.

- Taking Advantage of Existing Community Assets

Public policy and investments should focus on getting the most out of what we've already built. One of the largest problems with urban sprawl is the abandonment of buildings, transit systems, and other major investments. Reinvesting in our existing communities by maintaining parks, neighborhood schools, train stations, and historic buildings is central to smart growth's "fix it first" ethic. Environmentally minded developers and **community organizers** will be involved in revitalization efforts by working with or within city governments, local agencies, and other decision-making organizations to protect and reinvest in existing assets.

BOX 16-2

SMART GROWTH PRINCIPLES

- Mixing Land Uses
- Taking Advantage of Existing Community Assets
- Creating a Range of Housing Opportunities and Choices
- Fostering "Walkable," Close-Knit Neighborhoods
- Promoting Distinctive, Attractive Communities with a Strong Sense of Place
- Preserving Open Space, Farmland, Natural Beauty, and Critical Environmental Areas
- Strengthening and Encouraging Growth in Existing Communities
- Providing a Variety of Transportation Choices
- Making Development Decisions Predictable, Fair, and Cost-Effective
- Encouraging Community and Stakeholder Participation in Development Decisions

---

- Creating A Range of Housing Opportunities and Choices

  Not everyone wants the same thing. Communities should offer a range of options: big houses and small houses, condominiums and apartments, and "granny flats" for empty nesters. **Architects** and developers should be allowed to create a range of housing opportunities and choices, as they have immense influence on the design, promotion, and provision of all housing options, including affordable housing. These options should also include environmentally sensitive and effective building design, construction, and use of materials.

- Fostering "Walkable," Close-Knit Neighborhoods

  A compact, walkable neighborhood contributes to peoples' sense of community because neighbors are more likely to know each other, not just each other's

BROWNFIELDS SITE MANAGER

AT A GLANCE

Employment:

4,700

Demand:

Good

Breakdown:

Private sector, 60 percent

Public sector, 40 percent

Trends:

• Employment and work opportunities are subject to the rise and fall of the real estate market and shifting government regulations regarding brownfields sites.

• Professionals must adjust to increasingly complex and competitive applications for government subsidies and grants.

Salary:

Entry-level positions range from $50,000 to $60,000 (and are rare, given the qualifications required). Median annual incomes are around $70,000, and top salaries reach $120,000 or more.

JOB DESCRIPTION

Brownfields site managers' work makes thousands of acres of contaminated lands safe for redevelopment. "A brownfields site manager is a real estate developer who understands and handles all aspects of brownfields cleanup," says Stephen Villavaso, co-director of the Center for Brownfields Initiatives in New Orleans. "This work involves taking land that's already been developed and polluted, and reusing it. It ensures sustainability so that land is not wasted."

These professionals work primarily for private consulting firms or government agencies. They oversee the many steps involved in cleaning up polluted lands that do not fall under the Environmental Protection Agency's classification of Superfund sites. These cleanup activities include hiring specialists to test soil, water, and air samples for contamination; coordinating the excavation and transfer of polluted soil to landfills; removing or restoring deteriorated buildings or infrastructures; overseeing groundwater purification; and installing and monitoring water wells.

On the business side, brownfields site managers first identify land for redevelopment and then secure funding to revitalize the environmental and health quality of the property. These activities require professionals to negotiate real estate transactions and look for financial backing through grants, private donations, and other funding options. On the technical end, brownfields site managers then act as environmental engineers during land revitalization. They work with engineers and regulators to devise cleanup plans to ensure safety and prevent contamination.

The complexity of brownfields redevelopment demands that professionals work with local officials, lawyers, bankers, city planners, and neighborhood associations to complete restoration and development of contaminated sites, ensuring that redevelopment plans fit with a municipality's land use strategy and master plan.

Steven Santos, brownfields redevelopment coordinator for the Oregon Economic and Community Development Department, says his approach for dealing with communities is to hold informative meetings for residents to communicate the steps involved in cleanup plans. "I'm usually the primary contact for groups like nonprofits, cities, or towns that are seeking to assess and clean up brownfields sites," he says. "There's a lot of public speaking involved."

GETTING STARTED

• A bachelor's or master's degree in environmental management, land use planning, or urban planning, with coursework in real estate and business, will provide a good background.

• Working with real estate firms or financial institutions that specialize in dealing with distressed or contaminated land will give you an advantage.

• Working with community outreach programs will provide valuable project coordinating experience.

ADVICE FROM THE PROS

Santos suggests that aspiring brownfields site managers determine which aspect of brownfields cleanup intrigues them. "There are so many levels to this job," he says. "One way to figure out what interests you is to explore environmental studies or policy in college. Another is to attend a networking event, like the National Brownfields Conference." The conference website is www. brownfields2004.org.

cars. These communities offer safe places for people to walk and bike—sidewalks and trails—as well as things to walk and bike to, whether it's the corner store, transit stop, or school. Public health advocates and medical professionals promote walkable neighborhoods because they counter the sedentary lifestyles that sprawl encourages, which have led to higher levels of obesity, a disease that has become the nation's leading cause of premature death.

City planners, park and recreation employees, and local government officials should consider providing additional amenities for pedestrians and cyclists, including dedicated walking paths, bike lanes, bike racks, and other facilities. Such amenities will help fully realize the potential for "active living" communities that have well-connected street networks, intersections, and sidewalks. Ultimately, these employees who determine community layout and design are responsible for creating compact and walkable neighborhoods. **Architects**, developers, and planners involved in the "new urbanist" movement, which prioritizes designing neighborhoods for people rather than cars, champion such efforts.

- Promoting Distinctive, Attractive Communities with a Strong Sense of Place

Every community has landmarks that make it special, such as town squares, historic buildings, or neighborhood schools. These sites should be protected, rehabilitated, and celebrated. Unfortunately, unique landmarks are often terribly neglected, and many have been demolished. But that needn't be the case. In Washington, D.C., for example, Union Station is widely regarded as a masterpiece of American architecture. However, the train station was so neglected that by the 1970s, people called for its demolition, much like the razing of the original (and once beautiful) Pennsylvania Station in New York City. Rather than replacing Union Station with a miserable facility, congressional and transportation officials championed its rehabilitation and revitalization. Today, it is a remarkable economic development success story and popular destination and meeting place for travelers, local diners, moviegoers, and shoppers.

- Preserving Open Space, Farmland, Natural Beauty, and Critical Environmental Areas

Americans greatly value preserving open space and want access to natural areas. Even the National Association of Realtors has found that "when it comes to open space, voters are willing to support protection of most lands that are currently

327

undeveloped." Most "traditional" environmental professionals, including biologists, hydrologists, and geoscientists, are involved in preserving open space, farmland, natural beauty, and critical environmental areas. Also playing an important role in protecting these lands and preventing sprawl are **conservation biologists**, **environmental protection technicians**, **real estate associates** working for land trusts and natural resource agencies, **water resources managers** working for water management districts, and farmers working on agricultural land.

• Strengthening and Encouraging Growth in Existing Communities

Before paving over forests and farms, communities should look for opportunities to grow in already built-up areas. This approach reinforces the efforts of countless community development corporations, civic organizations, and affordable housing providers nationwide to create alternatives to sprawl and poorly planned development, roadways, and transportation infrastructure. Smart growth is particularly needed in America's lowest-income neighborhoods, where new investment or revitalization of any kind is extremely scarce, though much-needed to provide greater economic opportunities.

According to Harvard economist and business school professor Michael Porter, inner cities actually have some competitive advantages over sprawling communities because of their proximity to central markets and the availability of relatively low-wage workers. Yet relatively few businesses have taken advantage of these opportunities. Also, few cities recognize how existing cultural, architectural, and institutional amenities in their central cities could heighten quality of life. Carnegie Mellon University's Richard Florida contends that communities devoted to boosting quality of life will be more competitive than other places because they attract "talent workers" who are characteristically highly paid, childless, and free-spending—a very desirable demographic group.

• Providing a Variety of Transportation Choices

Most people love their cars, but resent getting stuck in traffic and being forced to drive everywhere. The public's tremendous frustration with traffic reflects its desperate desire for solutions. The Texas Transportation Institute recently listed "more travel options" as the first of seven recommended strategies for reducing traffic congestion. But people can't reduce their reliance on cars unless there are other ways to get around.

More communities need safe and reliable public transportation, sidewalks, and bike paths. Many transportation agencies are dedicating energy and resources to providing these alternatives, and the younger generation of traffic engineers and planners appears to be rising to the challenge. By 2010, the transportation field will turn over half of its personnel, creating new jobs at public transit and transportation agencies for those professionals interested in implementing alternative transportation choices.

- Making Development Decisions Predictable, Fair, and Cost-Effective

  Builders looking to implement smart growth should face no more obstacles than those contributing to sprawl, but this is not always the case. For example, higher-density infill development projects frequently face greater difficulty in securing financing, permits, and insurance merely because they are unusual and therefore perceived to be higher-risk projects. In reality, there's tremendous market demand for such projects, as "new urbanist" **architects**, planners, and developers have demonstrated time and again. Solutions include strategies such as rezoning areas to allow for mixed uses and higher densities where it's appropriate—in cities and business districts, for example, where compact urban form is desirable and compatible with what's already there. Some people also are focusing on providing incentives for smarter development. The U.S. Green Building Council, in fact, is developing a set of smart growth award guidelines to commend projects that represent green alternatives to sprawl. Again, such efforts involve local **elected officials**, planners, **architects**, and developers.

- Encouraging Community and Stakeholder Participation in Development Decisions

  Plans developed without strong citizen involvement don't have staying power. When people feel left out of important decisions, they won't be there to help out when public support is critical to a project's approval. Encouraging citizens and stakeholders is an important task for local government, as participatory decision making is absolutely critical for creating sustainable urban areas that serve all citizens. Local organizations, and especially **community organizers**, can help ensure the public has a voice in development decisions that effect their communities and living conditions. They can, for example, conduct community-planning meetings to develop long-term visions for their neighborhood or region. With long-term goals in mind, citizens will then be more capable of providing meaningful feedback about short-term development proposals.

LAND USE PLANNER

Employment:

33,000

Demand:

Excellent

Breakdown:

Public sector, 70 percent

Private sector, 30 percent

Trends:

• Increased emphasis on sustainable development will drive the demand for land use planners concerned with the environmental impacts of economic growth.

• Redevelopment and preservation of historic properties will increase employment demand.

Salary:

Entry-level salaries begin around $32,800, and median earnings equal $51,570. Most earn between $40,650 and $65,115, with those working in private firms earning top salaries that reach around $79,000.

JOB DESCRIPTION

Land use planners are creative thinkers who assess the social, economic, and environmental impacts of growth and seek alternatives to current development trends. Planners spend most of their time in the office reviewing project proposals, providing assistance to government boards, analyzing data, drafting contracts, and creating plans that promote smart growth and prevent sprawl. They must also prepare for, detect, and mitigate the effects of environmental intrusion. Planners frequently travel to building sites and meet with interest groups, citizens associations, and local officials to discuss the planning process. A politically charged occupation, planners must bring groups together over issues to work toward a sustainable future.

Some planners work for local, state, and federal governments, which makes them vulnerable to funding cuts. Others work for nonprofit organizations, private consulting firms, and industry, including coal and natural gas companies.

No matter where you work, "you have to help a community decide how to develop its space over time," says Rudayna Abdo, director of the American Institute of Certified Planners (AICP) and director of professional development for the American Planning Association. "You also must consider environmental regulations and policy issues as you plan a development."

GETTING STARTED

- A bachelor's degree in planning is required for employment. Pursue fields like architecture, geography, or social sciences.
- A master's degree is becoming essential for securing a job. Seek programs in urban or regional planning, public policy, natural resource management, or public administration.
- Programs accredited by the Planning Accreditation Board (www.acsp.org/pab) are highly regarded.
- Becoming a certified planner, through programs such as the American Institute of Certified Planners (www.planning.org/aicp), will give you an advantage.
- Strong skills in Geographic Information Systems (GIS), database systems, desktop publishing, and other computer programs are valuable.
- An understanding of government systems helps planners in their daily activities. Look for job and volunteer opportunities in local and state government offices.

ADVICE FROM THE PROS

Gwendolyn Hallsmith is director of Global Community Initiatives, a group that specializes in sustainable development, which is essential in the planning field. "People who work on sustainable development are countercultural and have to struggle to find resources for work in a world that seems, at times, to be committed to a path of self-destruction," she says.

With this in mind, Hallsmith says that keeping abreast of national and world news and learning about environmental issues has helped her to be successful in the land use planning field. "To do this work, you must understand the world in a different way than the corporate media and the dominant institutions in our society present it. Don't specialize in one narrow field of work, but instead, learn as much as you can about a wide variety of subjects and look for the interrelationships."

Career Spotlight

The City of Santa Monica
Santa Monica, California

"The ideas behind smart growth are good, but we can do even better," says Dean Kubani. "Let's imagine sustainable cities throughout the nation." One city that has gone beyond dreaming is Santa Monica, California, where Kubani is a senior environmental analyst and the sustainable city coordinator. The municipal government of this oceanside city has a full-time staff of 2,000 people in departments that include environment and public works management, planning and community development, police, fire, and more. There's a role for all of them in the sustainability effort—an effort that is built upon the principles of implementing smart growth within their community.

Santa Monica's road map is outlined in a "Sustainable City" plan and measured by over 100 indicators used to track ecological, economic, health, and community factors, all of which reflect smart growth principles. Development and city approval of such plans and indicators are important first steps that gauge a community's interest in and commitment to the hard work ahead. A handful of United States municipalities have taken this step.

Assigned to both the city's fifteen-person environmental programs division and the three-person sustainable city work team, Kubani oversees data collection and reporting against the Sustainable City plan. He manages programs that help businesses and residents adopt sustainability practices and ensures that the plan's objectives are incorporated into long-range planning efforts. He also directs a group that harnesses the city's purchasing dollars to support the plan. Most importantly, he is a tireless advocate who can be found speaking to groups around the city on almost any given day.

What kind of career path does a "Sustainable City Coordinator" take? "A circuitous one," Kubani laughs. He traveled to local government via science, consulting, and activism. After finishing a B.S. in geological sciences at the University of California–Santa Barbara, he spent two years in Australia, earning a master of applied science degree in engineering and environmental geology from the University of New South Wales in Sydney. The next six years were packed with new experiences. A brief

Dean Kubani (seated) and Karl Bruskotter examine land use patterns in Santa Monica.

career as a geologist with a soil/groundwater contamination consulting firm was followed by nonprofit stints at the Climate Institute in Washington, D.C., Natural Resources Defense Council, and Heal the Bay. Along the way, Kubani picked up knowledge of solid waste management, coastal water quality, energy efficiency, and municipal programs for reducing greenhouse gas emissions—all of which have served him well since coming to Santa Monica in 1994 and being promoted to his current position in 1999.

Environmental program analyst Karl Bruskotter joined Kubani's tiny team in 2001 and is practically a division by himself. "My work is focused on assuring that the city government itself uses sustainable practices in three areas—toxics use reduction, environmentally preferable purchasing, and integrated pest management," Bruskotter explains. "All three involve transforming a traditional practice that produces damage into a new one that doesn't, and that's incredibly rewarding."

Bruskotter's extensive resume includes environment, health, and safety executive positions at Air Liquide, an industrial and medical gas distribution company, and the global healthcare company Baxter Healthcare. He also has consulting experience at EOS Environmental, a wastewater treatment company that uses microbiology, biochemistry, and engineering in its approaches, and at the environmental engineering company McLaren/Hart. After obtaining both B.S. and M.S. degrees in environmental and occupational health from California State University at Northridge, he also worked for the Los Angeles Metro transit system.

An essential aspect of Santa Monica's path to sustainability requires changes to the built environment. Green building program advisor Greg Reitz heads the effort. "I have four main responsibilities," Reitz says. "Enforce current green building codes, suggest new ones, create programs that promote green building practices, and educate the public." The first task on the list involves checking all building plans against a Green Building Ordinance that is one of the best in the nation, with solid guidelines for energy efficiency and the use of recycled content in building materials. Reitz's B.S. degree in cybernetics from UCLA and a LEED (Leadership in Energy and Environmental Design) accreditation from the U.S. Green Building Council make him well prepared for the job.

Will Santa Monica become a "sustainable city"? Lasting change ultimately depends on the city's residents—so perhaps it's appropriate to give the last word to environmental outreach specialist Andrew Basmajian. "We'll use anything," he says. "Posters on buses. Teacher training. TV advertising. Workshops, brochures, retail displays, festivals. We want the people of Santa Monica to support the implementation of their Sustainable City plan." Accomplishing this goal will help the Sustainable City team create a sustainable community that clearly demonstrates how effective the smart growth alternative can be.

- Creating Livable Communities

Today, there's no shortage of places trying to implement smart growth. Among states, Massachusetts, New Jersey, Utah, Maryland, Michigan, Vermont, Oregon, Delaware, and many others have vigorous smart growth initiatives designed to curb urban sprawl, preserve open space, and reinvest in older communities. Metropolitan areas like Portland, Oregon, Salt Lake City, Utah, and Charlotte, North Carolina have also launched groundbreaking efforts to reshape the way they grow.

Leaders in these states recognize that quality of life strongly influences their economic competitiveness. According to experts on "the new economy," communities determined to improve quality of life and better manage growth are more likely to have a competitive advantage over those that take a laissez-faire or "growth for the sake of growth" approach. Real estate analysts concur. As stated in the annual industry publication *Emerging Trends in Real Estate 2002*, "Properties in better-planned, growth-constrained markets hold value better in down markets and appreciate more in up cycles. Areas with sensible zoning (integrating commercial, retail, and residential), parks, and street grids with sidewalks will age better than places oriented to disconnected cul-de-sac subdivisions and shopping strips, navigable only by car." Even the mainstream is beginning to recognize these factors. *Money* magazine recently ranked Portland, Oregon, as the best place to live in America, largely because of its growth management, public transit, and downtown revitalization policies.

Overall, the reason that places like Portland are so livable is because of their convenience, effective transportation systems, sound environmental quality, and nearby recreation facilities that give people the opportunity to enjoy the natural environment. Additionally, livable communities have relatively low levels of aggravation and poverty, and high levels of economic opportunity. They instill community identity through the preservation of their institutions, landmarks, and history.

Career Advice from the Expert

Aspiring professionals interested in pursuing smart growth strategies should consider a broad range of professions. Smart growth is not just an issue; it's a new and more thoughtful way to grow our communities. Accordingly, the success of smart growth relies on many professionals, from city planners and home builders to campaign finance reformers and traditional environmentalists. Informed advocates in all professions can influence this field.

Because smart growth requires interdisciplinary approaches and the joint efforts of multiple stakeholders, students should pursue a well-rounded education that includes economics, earth sciences, architecture, and political science. We tend to think holistically about growth and development issues, and typically address multiple issues simultaneously—such as the interaction between urban form and public health, and overall community quality of life in blocks, neighborhoods, and regions. Accordingly, aspiring smart growth advocates should seek training in problem solving that looks at multiple issues on multiple levels.

Other useful skills depend on the chosen profession from which someone approaches smart growth. At Smart Growth America, because we are a non-profit advocacy coalition, we look for people who have experience working on political campaigns and organizing. We greatly value the ability to absorb and analyze varied information, as well as the ability to use technical communications skills to reach target audiences.

Overall, the most effective smart growth professionals are those who understand the way the sprawl system works and can envision how various reforms will facilitate smarter development. Aspiring professionals should look for private sector experience in real estate, finance, or land development; public policy experience in housing, environmental issues, and transportation; or advocacy experience through grassroots organizing, political campaigns, or lobbying. Many of these skills can also be learned on the job.

RESOURCES

Brookings Institution Center on Urban and Metropolitan Policy,
 www.brookings.edu/urban

Congress for the New Urbanism, www.cnu.org

International City/County Management Association, www.icma.org

National Neighborhood Coalition, www.neighborhoodcoalition.org

National Trust for Historic Preservation, www.nthp.org

New Urban News, www.newurbannews.com

Santa Monica Sustainable City Program, http://santa-monica.org/epd/scp/

Smart Growth America, www.smartgrowthamerica.org

Smart Growth Network, www.smartgrowth.org

Surface Transportation Policy Project, www.transact.org

Urban Land Institute, www.uli.org

17. RECYCLING AND
SOLID WASTE

Editor Patricia Anne-Tom in her office at *Waste Age* magazine.

A CONVERSATION WITH
Patricia-Anne Tom

Reporting on garbage puts Patricia-Anne Tom right in the middle of the nation's mounting waste problem. "The amount of waste we produce is staggering," she says. "Fortunately, we have the tremendous opportunity to institutionalize recycling as a way to deal with it." Tom is concerned, however, that we may not grasp that opportunity. "Society and industry are responsible for continuing the previous successes from the past fifteen years. We must ensure that we don't literally throw them away."

Tom has been involved in the waste industry for more than five years, working on Primedia Business Magazines and Media's *Waste Age* and *Waste Age Product News* publications. As editor, she oversees both publications' editorial content and production. *Waste Age* has received more than twenty awards for editorial and design excellence during her tenure.

Prior to delving into the subject of garbage, Tom was associate editor for *Sports Trend* magazine. She has also worked for *Honolulu* magazine and as a Washington, D.C., correspondent for the *Billings Gazette*, *Montana Standard*, the *Missoulian*, and the *Helena Independent Record*. Tom received her M.S. degree from Northwestern University's Medill School of Journalism and her M.B.A. from Georgia State University.

What Is the Issue?

Humans continue to generate increasing amounts of trash, which must be managed and disposed of by the waste and recycling industry. According to the Environmental Protection Agency (EPA), the United States alone generated 229.2 million tons of trash in 2001. On average, Americans generate 4.4 pounds of garbage per person per day, which is about twice as much as we generated, per person, in 1960. Other industrialized countries face similar increases. Japan's waste generation rates, for example, increased 1.77% from 1999 to 2000. As lesser-developed countries become more industrialized, they'll likely generate more and more waste, too.

The increases are in part occurring because waste is built into the dynamics of our economy. Humans value convenience and rely on time-saving products and gadgets that make their lives easier. Additionally, so many products— CD players, razors, VCRs, and the like—are cheaper to dispose of than to repair, and so we tend to throw them away. Many countries, and the United States in particular, lack significant legislation or infrastructure to encourage waste reduction and recycling. Consequently, all of the waste generated must be managed; otherwise, it will create public health problems.

Management involves three basic steps: collection, processing, and disposal. First, waste must be collected from each residential doorstep and business. Once waste is collected, it must be processed—recycled, composted, burned at waste-to-energy plants, or consolidated with other waste. The third and final step is disposal, in which waste that is not recycled, composted, or burned is placed in a landfill.

While the waste management process appears to be relatively simple, each step has inherent problems and is highly regulated by regional, state, and federal governments. And management costs increase each time waste is handled. Let's look at each of these processes individually.

• Collection

Collection represents the most labor-intensive and expensive part of the waste management business. Every day, cities, counties, and private haulers stop at homes and businesses to pick up trash. These frequent stops contribute to wear and tear on vehicles as well as increased diesel fuel emissions. According to Inform, a New York–based environmental research firm, trash trucks have the lowest fuel efficiency of any vehicle type; on average they get only 2.8 miles per gallon. While heavy-duty diesel-powered vehicles, including refuse trucks, make up only 7 percent of vehicles on the road, they produce 69 percent of the on-road fine particulate pollution and 40 percent of the nation's nitrogen oxide emissions. New truck-engine regulations that will be phased in over the next five years are expected to reduce emissions. However, they will simultaneously increase vehicle cost and the amount of truck maintenance required; reduce the fuel economy; and decrease engine horsepower, meaning that haulers will spend more time and money collecting trash.

As communities look to divert more waste from landfills, the types of waste that must be collected and separated for recycling further complicate collection. Many towns, for example, prohibit leaves and grass from being thrown out with regular trash so that they can be composted. In some communities, food waste, traditional recyclables (aluminum, glass, and paper), and electronic waste (computers, cellular phones, and televisions) cannot be thrown out with regular trash. The more types of materials that must be separated from garbage, the higher haulers' costs.

Finally, finding, training, and retaining people to work in the waste industry complicates collection. Collecting trash is not a glamorous job. The work is labor-intensive, which makes recruitment difficult and safety a concern. Recent increases in health and insurance premiums have heightened pressures on collection.

• Processing

Once waste and recyclables are collected, they are sometimes processed or recycled before heading to disposal sites. For example, waste often is taken to a transfer station, where it is consolidated with other communities' waste before it is taken to a landfill. Or waste can be recycled, composted, or burned at waste-to-energy plants. Communities are striving to "recover" more from the waste stream for beneficial use due to fewer landfills in further-distanced locations, increasing difficulty in finding new areas to construct additional disposal sites, and a growing concern about environmental consequences. The EPA states that the recovery rate, including composting, increased from approximately 28 percent in 1999 to approximately 30 percent in 2001. Many communities set their waste diversion/recycling goals at 50 percent or greater. However, the United

States lacks legislation that encourages manufacturers to develop products that are easily recyclable.

As a result, finding the most efficient system to separate, process, and market materials is an issue. It's one thing to want to conserve resources to preserve the environment, but waste managers also have to stay in business. So they must continually evaluate their recycling programs to determine which materials make economic sense to collect and recycle. The industry must also continually develop new markets for recyclables, such as using glass as roadbed aggregate or old tires to make artificial turf. Ultimately, recycling must be profitable to survive.

• Disposal

As a last alternative, waste winds up in landfills. In developing countries, landfills sometimes are unsafe and subject to landslides because of poorly engineered waste placement methods. People frequently scavenge these sites, further contributing to safety issues.

Developed countries have stringent regulations governing landfills to make sites safer. U.S. landfills, for example, are regulated by federal and local authorities to ensure that liquids and gases emitted do not harm human health or the environment. Landfill owners not only have to follow daily regulations to operate a landfill, but they must also show that they have the financial capacity to manage sites thirty years after they close.

These regulations have increased operator costs, and that has decreased the number of landfills operating today. In America, four out of every five landfills that were operating in 1984 have since closed. Moreover, no one wants a landfill in his or her community, which makes new landfills difficult to site. This has left approximately 1,858 landfills operating nationwide. Again, today's landfills are larger and farther apart, meaning waste must travel greater distances to reach disposal sites.

How Are Environmental Professionals Approaching the Issue?

Most people tend to forget about garbage once they've tossed it into their trashcans, so they don't realize how much trash they generate or understand the problems inherent in managing waste and recycling. Reducing waste generation rates will always be a difficult proposition. It requires educating consumers about garbage and its effects, and then changing their consumption and disposal habits.

- Improving Collection Operations

Although collection is the least sophisticated aspect of the waste management business, waste haulers are making improvements and becoming more efficient. Particularly with today's equipment, environmental professionals are addressing the problems associated with high collection costs and intensive labor requirements.

For instance, automated trucks now frequently service collection routes, reducing or eliminating the driver's need to get out of the vehicle to empty trash cans. This has reduced both the number of driver injuries and the number of workers required to manage collection. Additionally, manufacturers have improved equipment to ease collection labor. Trucks now provide drivers with more cab room and easy-to-operate controls to make operating environments more comfortable.

Collection managers also are using technology to improve operational efficiency. For example, global positioning systems and routing software have streamlined routes in many communities. According to Waste Industries, routing software can help improve productivity by 8 percent in some instances.

- Evaluating Recycling and Enacting Waste Diversion Policies

In the processing part of the waste business, professionals continually evaluate waste generation and disposal practices to determine the most efficient system for processing and marketing recyclables. They also seek to improve trans-

BOX 17-1

STRATEGIES TO PROMOTE RECYCLING AND REDUCE WASTE

- Improving Collection Operations
- Evaluating Recycling and Enacting Waste Diversion Policies
- Targeting Specific Waste Streams
- Creating Disposal Alternatives
- Educating Waste Generators

portation networks to move waste to disposal sites. Recycling's success hinges on the ability of waste management professionals' **recycling coordinators** at local government agencies and their contractors to develop programs that are more than just good for the environment; the programs must also pay for the cost of collecting and processing recyclables. Consequently, **recycling coordinators** working for municipalities, counties, and private firms are examining recycling program best practices, what materials to collect, and how high to set goals for waste diversion—the conscious redirection of waste toward recycling and away from landfills. Industry leaders like the National Recycling Coalition and regulators like the U.S. EPA are encouraging product manufacturers to be more proactive in developing products that can be recycled and to think about their products' afterlife.

In 2002, the National Recycling Coalition conducted a study to define recycling's presence in America, which helped to emphasize its importance in the industry. The resulting report indicated that recycling employs more than 1.1 million people, who are paid $37 billion annually. It also showed that recycling generates revenues equal to the auto and truck manufacturing industries, with gross annual sales of $236 billion.

Not one to rest on the laurels of these successes, the Solid Waste Management Association of North America (SWANA)—a trade association that represents private and government sector waste management professionals—recently issued a strategy to increase waste recovery and reduction efforts. The plan, "Pushing the Envelope on Waste Reduction and Recovery," makes six policy recommendations to remove barriers to higher diversion:

1. Encourage more extensive product stewardship or the improvement of a product's environmental and social performance—usually by assessing its entire life cycle—by product designers, manufacturers, retailers, and consumers.
2. Expand efforts by federal, state, and provincial governments to develop markets for recycled materials and recovered energy.
3. Provide financial incentives for investments in recycling, composting, and the use of recycled materials.
4. Include waste-to-energy and conversion technologies in renewable portfolio standards—a requirement that all electricity providers include a minimum percentage of renewable energy sources in the electric power supply options

RECYCLING COORDINATOR

AT A GLANCE

Employment:

15,000

Demand:

Average growth

Breakdown:

Public sector, 60 percent

Private sector, 40 percent

Trends:

• While southern states are seeing a surge in recycling programs, many northern and coastal states with well-established recycling programs are cutting back on funding and operating with smaller workforces.

• In the next few years, more job openings will result from increases in population and the amount of waste generated per person, as well as increases in disposal requirements and in the amount of refuse recycled.

• Competition will be keen, as recycling companies and municipalities continue to shed excess layers of management to increase productivity and competitiveness.

Salary:

Entry-level salaries start around $30,000 to $35,000, with median salaries reaching $50,000 and top wages over $75,000.

JOB DESCRIPTION

Recycling coordinators work for municipal governments or private firms primarily to supervise an estimated 13,000 curbside recycling programs and 5,000 drop-off programs in the country. They design and implement solid and hazardous waste management programs for local cities, towns, and other communities. They also communicate the environmental advantages of recycling materials through public outreach campaigns. "These professionals are really on the front lines of the recycling battle," says Michael Alexander, senior policy associate at the National Recycling Coalition (www.nrc-recycle.org). "A talented, motivated person can really have a big impact on a city or town."

Kara Dinhoffer, recycling coordinator for the city of Boulder, Colorado, says she developed Boulder's recycling plan from conversations with community and city council members. As she creates annual budgets, Dinhoffer aims to increase waste reduction within financial constraints, while delivering the best recycling program possible.

Recycling coordinators must also manage compliance with town or city ordinances, coordinate collection schedules, manage contracts with private waste management firms, apply for grants, and confront any other issues that may arise in a recycling program. Many recycling coordinators work for local government, sometimes serving one city or town, while other times working with a number of different municipalities. Numerous professionals work for private firms under contract to towns and cities. Recycling coordinators can also work in the private sector, promoting and coordinating in-house recycling and waste reduction programs for large companies.

GETTING STARTED

• While there is no recommended course of study, a bachelor's degree is required for employment. Some colleges offer waste management programs, but recycling coordinators can hold degrees in a variety of subjects.

• Volunteer or fieldwork for local recycling programs or private waste management companies will give you an advantage in finding employment.

• Budget and project management experience can be helpful, as recycling coordinators are responsible for creating an annual budget and overseeing community-wide projects.

ADVICE FROM THE PROS

Jan Ameen, executive director of the solid waste district in Franklin County, Massachusetts, says her work involves more than just recycling waste products. "We're involved in every aspect, especially things like selling compost bins, and knowing which materials go where, and coordinating trucking routes," she says. "It's a tricky field that demands 'seat-of-the-pants learning.'" Ameen recommends that newcomers explore internships in their city or town and progress from there. "Everyone I know in this field has worked their way up the ladder. It's how this field is."

they offer their customers (such as wind, solar, biomass, and geothermal energy). Also include green power programs.

5. Encourage the recovery and use of landfill gas by reinstating federal tax credits and through renewable portfolio standards and green power programs.

6. Support technology transfer and research and development efforts that have the potential to significantly increase waste recovery rates and work to reduce barriers to their implementation.

The EPA is also encouraging communities and businesses to divert more of their waste by setting new recycling goals. In 2003, it issued its Resource Conservation Challenge to help boost the national recycling rate from 30 to 35 percent, and to cut the generation of thirty harmful chemicals by 2005 through public awareness about existing electronics recycling programs (sponsored by several corporations).

• Targeting Specific Waste Streams

In addition to enacting operational and policy changes to foster efficiency, communities are examining their waste diversion strategies, identifying specific waste streams for recycling, or considering incentives to encourage the recycling of specific products. Many communities have implemented yard waste bans and separate green waste collection programs, which has led to yard waste accounting for half of the total waste reduction in the year 2000.

More than 6,000 communities nationwide are attempting to connect consumers' disposal habits to their pocketbooks and educate people about waste and recycling by implementing pay-as-you-throw (PAYT) programs. PAYT programs are different from typical flat-rate household waste disposal fee programs. With PAYT, residents' disposal bills are based on the amount of trash they throw away; the more they dispose of, the higher their garbage bills. Simultaneously, because residents are not charged for the amount they recycle, this creates waste diversion incentives.

San Francisco recently targeted a specific waste stream—food waste—by implementing a program in which food waste is collected separately and then processed into compost. The finished compost is sold to area farmers, whose products are then sold to city restaurants. By diverting food waste from landfills, San Francisco hopes to reach its 50 percent waste diversion goal. By targeting specific waste streams through policy changes or implementing new recycling programs, communities believe they will get closer to achieving their diversion goals.

The industry is also encouraging manufacturers to take more responsibility for the waste they generate and to pay for recycling. Environmentalists, in particular, are encouraging electronic equipment manufacturers to take back obsolete equipment and manage waste internally, as the EPA estimates 500 million computers are expected to become obsolete by 2007. In 2003, California became the first state to pass an electronic waste bill that mandates consumers to pay an electronic products fee to cover the costs of recycling electronic devices.

• Creating Disposal Alternatives

Disposal options are improving as well. As noted, the operation of U.S. landfills is costly because of heavy governmental and engineering requirements applied to these facilities. This has led owners and operators to continually seek ways to better manage their disposal resources. For example, to stay up-to-date on regulations and best waste disposal techniques, many states require landfill operators to become certified.

To determine whether they can increase existing landfill value, operators currently are investigating bioreactor landfills, in which liquids are added to waste to speed up decomposition processes and allow more room for waste. The EPA recently reviewed its landfill regulations to address research, development, and demonstration permits that would allow states to grant exemptions to the federal laws for landfill research programs, including bioreactor projects. Industry professionals are hopeful that the results of these pilot projects will improve landfill management nationwide.

Owners and operators also continue to explore other avenues that can create environmental and financial benefits. For example, closed landfills sometimes are converted into golf courses or parks for public use. The EPA's Landfill Methane Outreach Program (LMOP) encourages manufacturing facilities to convert landfill gas to energy, which creates voluntary partnerships between landfill operators and **environmental engineers** to use waste as a renewable energy source. As of June 2003, more than 340 landfill gas (LFG) energy projects were operational nationwide. The EPA estimates that more than 600 other landfills present opportunities for LFG project development.

Landfill experts also continue to educate themselves and learn from their

ENVIRONMENTAL CONSULTANT

AT A GLANCE

Employment:

96,000

Demand:

Slow

Breakdown:

Private sector, 100 percent

Trends:

• Environmental consulting is rapidly dividing into small firms with narrow geographic or niche focuses, and very large national and international firms that cover wide-ranging environmental work from design to completion.

• The consulting industry continues to experience mergers and acquisitions, resulting in fewer industry leaders dominating employment opportunities.

Salary:

Consulting firms hire from most of the forty job categories highlighted in this book. See individual jobs for salary information.

JOB DESCRIPTION

Environmental consultants offer expert advice to local, state, and federal government agencies and private-sector clients who need to adopt environmentally sound practices or clean up contaminated sites. "We assist clients in implementing and adhering to environmental regulations for compliance purposes," says Richard Moore, environmental consultant manager for Trinity Consultants in New Jersey.

The bulk of billions spent annually for environmental consulting goes toward water and wastewater management, solid and hazardous waste management, and the regulatory closure and remediation of contaminated sites—both on and off the federal government's Superfund list. However, other consultants also work on risk assessment, brownfields redevelopment, environmental impact assessments, capital project permitting, and environment, health, and safety management and compliance issues.

Some environmental consultants, such as Moore's group, are developing strategies for clients to improve air quality or design solid waste and recycling programs. They often look at ecological impacts of development and recommend solutions to mitigate those impacts on plants and animals, especially endangered or threatened species.

Environmental consultants usually come from scientific or engineering backgrounds, but are quick to point out that overspecialization can limit their scope of projects. Gary Floyd, environmental consultant for Tetra Tech in San Francisco, says the field demands quick, efficient work within a client's time and budget constraints. On any given day, Floyd meets with current and prospective clients, conducts site visits to determine the severity of a project, and proposes project cost, design, and implementation. He also evaluates results from environmental impact assessments, air, water, and soil sampling, and other lab data, in addition to documenting project progress.

Overall, this results-driven field is extremely fast-paced and competitive, with lots of time on the road and in the field. Client satisfaction is ultimately the measure of a consultant's success, and the best professionals are those who are confident and competent in their work, thereby inspiring trust in their clients.

GETTING STARTED

• A bachelor's degree in environmental management, environmental or chemical engineering, natural sciences, or another related field is required for employment.
• Look for work or internship experience at private consulting firms. The University of Michigan's School of Public Health (www.sph.umich.edu) offers an extensive list of private companies that offer experience for recent graduates.
• Networking through trade organizations, such as the American Society of Civil Engineers (www.asce.org), may help you locate work after graduation.

ADVICE FROM THE PROS

Two things that prove invaluable to consultant Richard Moore are his knowledge of computer programs and Geographic Information Systems. "By being proficient in computer software and calculations, you prove your capabilities to your clients and employer," he says, as mastery of sophisticated technology is a growing aspect of the consultant's world in order to serve clients faster, better, and cheaper. "Business and people skills also help you relate to your clients and provide them with valuable services."

peers via training seminars, conferences, and trade shows. To improve the safety of disposal abroad, industry professionals from the United States and other countries often lend their expertise to developing nations. Environmental professionals have also become more proactive about ensuring that waste generated in America is not improperly disposed of abroad.

- Educating Waste Generators

Finally, many in the industry are striving to educate primary waste generators —consumers and businesses. By doing this, we can begin to create an informed public who can reduce waste. Several landfills and recycling facilities, for instance, incorporate community education centers to explain where garbage goes. Waste managers often work with their communities when trying to implement a new waste program or site a facility. State departments of environmental protection generate awareness of waste and recycling issues through activities like America Recycles Day and cleanup events. And educators, called "garbologists," attempt to teach the public about waste management and recycling at colleges and universities.

Career Advice from the Expert

Relatively few people understand what happens to waste and recyclables once they are thrown away. Some universities, such as the Yale School of Forestry and Environmental Studies and the University of Wisconsin–Madison's Department of Engineering and Professional Development, offer courses on waste management and recycling, which can help industry newcomers get their foothold.

Many organizations are available to provide advice and, in some cases, financial assistance to solve industry problems. Prospective recycling and waste management professionals would do well to take advantage of these kinds of programs and organizations when beginning their job search or gaining preliminary experience. For instance, the Environmental Research and Education Foundation (EREF) attempts to answer challenging waste management questions by funding research projects and several scholarships to benefit the industry and the environment. Currently, Michigan State University is using a $500,000 EREF grant to design and operate a bioreactor landfill cell.

Educational, networking, and career advancement opportunities also are

Career Spotlight

Recycle America Alliance (RAA)
Houston, Texas

The same economic pressures that create eye-popping mergers in fields like banking, electronics, and energy also drive recycling companies to get bigger in an effort to serve customers better, increase revenues, and drive down costs. It comes as little surprise that industry leader Waste Management created the Recycle America Alliance (RAA) in 2001 as part of a strategic effort to improve economies of scale and profitability in the rapidly changing recycling marketplace. Waste Management owns 90 percent of Recycle America Alliance.

Before creating the alliance, Waste Management was already by far the largest collector of recyclable materials in North America, processing over five million tons of recyclables per year. The first alliance partner was Milwaukee-based Pelz Group, the country's largest privately held recycler. Combining with Pelz brought an additional two million tons of recyclable fiber annually into the RAA.

With 3,300 employees, RAA handles over eight million tons per year and is growing. It operates eight recycling plants and provides marketing for over 190 locations in the United States and Canada—where its distinctive trucks and dumpsters are a recognized part of the urban landscape. RAA also operates seven container processing facilities, one

Technicians at Recycle America Alliance dismantle computer hard drives as part of the electronics recycling program.

plastics recycling facility, and four electronics recycling facilities.

Kevin McCarthy, director of electronics recycling (E-Scrap) for RAA, observes, "Initially focusing on recycling bottles, cans, and papers from curbside and commercial collection, we now operate glass and plastics processing plants and handle electronics. It's an exciting, ever-changing environment, with new career paths and challenges."

Every day, RAA collects recyclables such as paper, glass, metals, plastics, and compost, and takes them to a material recovery facility for weighing, processing, and marketing. To ensure steady demand for recyclable commodities, RAA identifies buyers worldwide for postconsumer and postindustrial commodities such as fiber, nonfiber, scrap metal, textiles, rubber, electronic scrap, and plastics. To reduce commodity price risk, RAA placed 75 percent of its commodities under long-term floor price contracts and entered into fixed-price hedge contracts.

Mary Wendell represents RAA's entire range of services to a current base of 1,000 customers. Previously, she was founder and general manager of Data Destruction, a shredding company. As an RAA account manager, she spends time on customer maintenance and developing new business.

Wendell describes a typical sales sequence: "First, I tour a facility, performing a waste audit to reduce recyclables going into the waste stream. At the same time, I identify opportunities for cross-selling product lines, such as E-Scrap, shredding, trash collection, and equipment."

Once she has identified an opportunity, "educating businesses on recycling potential and employee participation is next. It's vitally important to have upper management support a recycling program. Fortunately, business leaders today understand that recycling is both good for the environment and good for business."

Matthew Coz is responsible for delivering on the sales Wendell closes. Coz manages nineteen facilities as director of recycling for the East. His career in the environmental field began in 1989 as a business development/proposal specialist. "This was an excellent starting point," he suggests, "allowing interaction with engineering, sales, finance, and operations.

"Every successful operation requires a strong team, so I work with my staff and develop their skills. We look at everything together, from engineering ideas to personnel scheduling and management, so each facility operates as safely and efficiently as possible." With a B.A. in psychology from St. Anselm's College and a master's in industrial/organizational psychology from the University of New Haven, he emphasizes that a recycling career today "requires a multidisciplinary approach to achieve success."

Kevin McCarthy would agree. "My career with Waste Management has evolved as the recycling industry has grown and diversified over the last ten years." Today, McCarthy reviews and sets financial and budget targets, sells to new customers, establishes productivity goals, and serves as a liaison with regulators and key interest groups that affect RAA's business.

McCarthy has a B.S. in environmental policy analysis and planning from the University of California–Davis. He acknowledges, "I was fortunate that the year I interned at the Capitol [in Sacramento] coincided with California passing its recycling law. Public demand created opportunity for me then and thousands of new jobs in the recycling industry—just as it continues to do today."

available at the industry's major trade shows, WasteExpo and WASTECON. WasteExpo is produced by Primedia Business Magazines and Media and co-sponsored by *Waste Age* and the Environmental Industry Association (EIA). WASTECON is produced by SWANA. Both trade shows feature equipment displays and educational seminars.

These organizations also represent a wide range of waste management and recycling professionals who can provide advice and expertise. The EIA, through its subassociations, the National Solid Wastes Management Association (NSWMA) and the Waste Equipment Technology Association (WASTEC), represents companies and individuals who manage solid and medical wastes; manufacture and distribute waste equipment; and provide environmental management, consulting, and pollution prevention-related services. SWANA provides education and training through conferences, certification programs, publications, and technical training courses.

Despite these offerings, the only way for students and recent graduates to truly understand the inner workings of the waste and recycling industry is to seek hands-on experience. During the past ten years, environmental professionals' level of experience in the industry has increased greatly, and they are much more sophisticated about operating profitable businesses. Thus, professionals wanting to make a difference in this industry would do well to learn as much as they can on the job.

Waste management is deceptively simple. So don't be discouraged if you find the learning curve is longer than a year. The good news is that time invested in the industry will be rewarding and well spent. Waste management and recycling professionals realize they can have a measurable influence on their communities just by looking at their curbsides. Moreover, for those who worry about job security, this industry is fortunate that there will always be waste and recycling.

RESOURCES

Environmental Defense, www.environmentaldefense.org
Environmental Industry Associations, www.envasns.org
Environmental Protection Agency Office of Solid Waste, www.epa.gov/osw
Environmental Research and Education Foundation, www.erefdn.org
Grassroots Recycling Network, www.grrn.org
Inform, www.informinc.org

National Recycling Coalition, www.nrc-recycle.org
Natural Resources Defense Council, www.nrdc.org
Recycle America Alliance, http://www.recycleamericaalliance.com
Solid Waste Association of North America (SWANA), www.swana.org
Waste Age, www.wasteage.com

18. SUSTAINABLE CONSUMPTION AND GREEN LIFESTYLES

Center for a New American Dream president Betsy Taylor working in her office. Source: Center for a New American Dream.

A CONVERSATION WITH
Betsy Taylor

"The incessant chase for more has staggering effects on the environment and our quality of life," says Betsy Taylor. "The planet cannot survive six billion people consuming as Americans do. Even if it could, our material definition of happiness leaves people yearning for more time to spend with their families, relax, and have fun."

It's no wonder Taylor is committed to these ideals. As president of the Center for a New American Dream, she works to help Americans consume responsibly in order to protect the environment, enhance quality of life, and promote social justice.

Before founding the center, Taylor was the executive director of the Merck Family Fund and vice chair of the Environmental Grantmakers Association. She received her M.P.A. from Harvard University and has written numerous op-eds. Recently she wrote *What Kids Really Want That Money Can't Buy* (Warner Books, 2003) and co-edited *Sustainable Planet: Solutions for the Twenty-First Century* (Beacon Press, 2002).

What Is the Issue?

Consumerism in American society is arguably more dominant now than at any time in history. Our most popular national pastime is watching television, followed closely by recreational shopping. The United States has the highest per capita consumption rate in the industrial world. While our material gains have improved quality of life in some notable ways, there are many hidden costs to our "more is better" definition of the American dream. Namely, unsustainable consumption damages the environment by depleting crucial resources such as water, fish, and trees, and generates astronomical amounts of solid and atmospheric waste.

Our hectic "work and spend" way of life also takes its toll on our financial well-being, psychological health, and personal happiness. In 2002, more than 1.6 million American families declared personal bankruptcy. American workers, on average, worked longer hours than citizens in any other industrial nation. Ultimately, the "more is better" approach to life widens the growing gap between the "haves" and "have-nots." Many people fear we are creating a fortress world where only the strong have access to necessary material resources.

• The Growing Gap Between the Rich and Poor

In today's world, 20 percent of the population consumes 86 percent of material goods, whereas the poorest 20 percent consumes only 1.3 percent. Almost half of the world's people live at the poverty level or below, living on less than two dollars per day. There is something deeply wrong with our world when trade, tax, and social policies permit so few to acquire so much, while billions of people scramble everyday merely to survive. Our global and domestic culture and economy help propel those at the top to consume more, while those at the bottom are continually left out. Further, the environment is inadvertently degraded throughout this process.

• Global Environmental Collapse

The ecological footprint is a measurement showing how current consumption and waste production exceed the earth's capacity to create new resources and absorb waste. For an explanation of how the ecological footprint is calculated, see chapter 7 on economics. The average American's footprint is almost five times above the world average and nine times above what the earth can actually sustain. Americans consume more fuel, aluminum, energy, paper, and meat per capita than any other society on the planet, even though we make up less than 5 percent of the world's population. Our consumerist culture is being exported all over the planet through globalization and increasing integration of the world's societies. If everyone on the earth consumed as much as the average American, scientists estimate we would need at least four additional planets to provide the necessary resources and absorb the resulting waste. The American economy and culture are simply not sustainable.

- ## Are We Happy?

The nightly news often focuses on U.S. productivity and consumer expenditures, yet rarely chronicles the environmental or social costs associated with our working and spending norms. Americans work longer hours than Australians, Japanese, and most Europeans. We take less vacation time than we did twenty years ago. People spend more time in front of electronic screens or in malls and less time hanging out with friends or walking in the woods. Our children also go to school for longer hours and do more homework. Why is all of this happening? In part, our long workdays, weeks, and years are connected to material consumption. Everything in our lives is getting bigger—cars, houses, television screens, muffins, waistlines, credit card debt, and so on. Many Americans are forced to work harder to pay off these bills.

No doubt, material goods have a place in the good life. They can provide comfort, leisure, and serve various real needs. But many people are beginning to slow down and ask, "What really matters?" Are we happier, as a nation, with our storage units, world-class malls, and revolving credit lines? Evidence suggests that after certain basic needs are met, citizens with far more material goods and wealth are not happier than those much further down the economic ladder. Instead, this culture is fostering epidemics of depression, stress, and insomnia. As social psychologist David G. Myers notes, since 1957 the number of Americans stating they are "very happy" has declined from 35 to 32 percent. Meanwhile, divorce rates have doubled, teen suicide rates have nearly tripled, violent crime rates have nearly quadrupled, and more people than ever (especially teens and young adults) are depressed.

Part of the race for more money is fueled by fear of having to go through life totally alone. In the United States, unlike all other industrial nations, young Americans have little help with housing, healthcare, college tuition, or other social safety nets. This isolation, exacerbated by American social policies, makes us that much more vulnerable to pervasive advertising that suggests we will find safety, happiness, and love in material goods. In this fast-paced, commercial world, people are increasingly yearning for balance, silence, and reconnection to nature. They want more fun with less stuff, as well as time for relationships, community, and more community service. For these reasons, many Americans are beginning to challenge the culture of excess consumption.

How Are Environmental Professionals Approaching the Issue?

Many scientists, engineers, educators, **policy analysts**, business leaders, and activists are working to fashion a sustainable economy, culture, and society. Some are focused on policy reform; others are mounting market-based campaigns to aggregate consumer demand in support of green products and companies. Some are educating children and college students about consumption as a root cause of ecological degradation, while others are involved in new technologies that will greatly reduce the material throughput involved with producing particular goods. The need for professional engagement in all sectors of society is of paramount importance.

• Changing Policy

Government-led change can be the most effective tool for combating unsustainable consumption because policy, rather than voluntary actions where not everyone will participate, is often the best way to enforce large-scale societal changes. Economists and **policy analysts** are designing new ways to encour-

BOX 18-1

CREATING A SUSTAINABLE SOCIETY

- • Changing Policy
- • Mobilizing Politically Engaged Citizens
- • Galvanizing Business Leaders
- • Reforming Businesses
- • Creating a Cultural Renaissance

age resource conservation, including plans for eco-taxation, eco-labeling, product take-back, and subsidy shifts. Many European nations and New Zealand have begun to implement these schemes.

Eco-Taxation

The United States is the only modern industrialized country without a significant energy pricing system, where the government fairly assesses and sets energy prices based on the natural resources such energy uses up. The federal government has consistently skirted this critical issue. Americans pay far too little for gasoline and oil, consequently squandering these resources, especially in our transportation and agricultural sectors. Political will is needed to implement tax policy changes such as nonregressive energy and natural resource taxes, like those that exist in Germany, for example. These changes must be done in a way that lessens potentially regressive outcomes and brings energy and other resource use in line with environmental limits. As with all tax policy work, such changes require skilled **lawyers** and finance experts to craft policy, committed advocates to create public pressure, and courageous political leaders to enact needed legislation.

Eco-Labeling

The United States is also one of the only nations that has fought aggressively against eco-labeling. Eco-labels provide information about the origins, production practices, and conditions under which products are created, in order for consumers to make wise purchasing choices. The Canadian, German, Swedish, and Dutch governments have each played a fundamental role in supporting legislation or programs that define sustainable products, certification methods, and available information. Because industry has fought against a government-backed label, the nonprofit sector is now developing eco-certification initiatives that are on par with other countries' programs for products like organic foods and fair trade coffee. The best known labels include the Forest Stewardship Council, which certifies sustainable forest products; the Marine Stewardship Council, which certifies well-managed fisheries and seafood products; and Green Seal, which certifies a host of products including cleaners, food packaging, paints, lodging properties, and water-efficient fixtures. Certification specialists working for these and other independent organizations are in charge of evaluating products to ensure they meet sustainability criteria before they can receive actual labels and certification.

Communications professionals including **green marketers** and public educators inform the public about green products and also help them understand how to use green labels in making responsible purchasing decisions.

Product Take-Back

Product take-back is an innovative policy mandating that industries be responsible for the ultimate disposal of the products they produce. Passing this kind of legislation changes industries' entire incentive structure, typically leading to the redesign of everything from cars to refrigerators in order to reclaim and reuse all valuable materials, rather than simply throwing them away after use.

Shifting Subsidies

Government spending and tax policy should help our nation become more sustainable. Instead, most federal subsidies go toward high-impact, dinosaur technologies, such as those used by the oil, gas, nuclear power, and coal industries—technologies that are outdated and create vast degradation. Shifting taxpayer subsidies to new technologies, such as wind and solar energy generation, will reward industries that are conserving resources, minimizing waste, and generating new jobs.

Building Political Will

The big work is political. There are many policy solutions, from taxing carbon emissions and creating eco-labels to major public investment in new agricultural, transit, and electrical generation technologies. Sadly, since 1980, the United States has lacked the political leadership to implement such strategies domestically. To do so, we must build public awareness and support those **elected officials** who grasp the urgent need to change direction. We must align our economic development and consumption models with the earth's capacity to sustain a healthy environment. **Community organizers**, communications professionals, and educators are vital in this effort.

- Mobilizing Politically Engaged Citizens

Mobilizing citizens through public education and advocacy campaigns is a key part of the solution. We don't need everybody to begin to shift the system. Sometimes it's easy to think it's hopeless, that too many Americans just won't

ELECTED OFFICIAL

AT A GLANCE

Employment:

30,000

Demand:

Few new elected offices are being created.

Breakdown:

Public sector, 100 percent

Trends:

- Many elections are uncontested.
- Nine out of ten officials work in local government.
- Fundraising abilities are increasingly important for success.

Salary:

Salary ranges vary greatly, from under $10,000 to nearly $200,000. In state positions, officials earn from as little as $60,000 in Arkansas to as much as $150,000 in New York. At the national level, congressional lawmakers earn from $136,700 to $175,000.

JOB DESCRIPTION

Elected officials are vital to the environmental movement. Will Toor, mayor of Boulder, Colorado, believes that the fate of the environment rests on decisions made by elected officials. Those working on pro-environment agendas in areas such as land preservation, pollution reduction, and environmental policy advancement, for example, draw attention to environmental problems that communities face across the country.

Elected officials work at the local, state, and federal level as city or town councilors, mayors, county commissioners, state representatives, state senators, congressional lawmakers, and even President of the United States. For some, serving in public office is a full-time paying career. Others work on a part-time or volunteer basis.

On the legislative end, elected officials introduce, lobby for, and vote on bills. This may include implementing laws at the state level for determining specific land uses, for example. At the federal level, elected officials pass legislation that affects the entire nation. First and foremost, however, officials make decisions based upon the needs and demands of their constituency. They also build and maintain relationships with constituents to represent them effectively at meetings, public hearings, and public outreach. For these reasons, they must have excellent public speaking skills and the ability to negotiate with other officials. They must also be respectful of differing views.

GETTING STARTED

• Elected officials come from all career paths. A bachelor's degree or further education often helps secure a public position. Some officials hold degrees in fields like biology, history, law, and English.

• Knowledge of government and legal systems is necessary.

• Working in student and local government is a great way to gain experience.

• Those aspiring to federal-level positions should seek volunteer or internship work for state and federal lawmakers.

ADVICE FROM THE PROS

Councilwoman Jane Conway, who has served the town of Smithtown, New York, for fourteen years, is known for her commitment to preserving open space and focusing on cleanup efforts. "Protecting the environment is good business practice on a local level," says Conway, "because it maintains the property values in a community and makes people proud of where they live." Conway advocates a "thick skin and strong stomach" for serving in public office. "You have to be prepared to put a lot of time and effort in this work. You must go into it with the idea that you're going to do your best, but it's time-consuming and you give up certain aspects of your personal life."

ever change. Realistically, a majority of Americans are highly unlikely to change how they consume, and companies won't change production practices unless the laws governing consumer and producer behavior require those changes. Rule changes almost always come from intensive public pressure that offsets the influence of huge corporate interests.

People need to know that what they do right now matters. Not doing anything can be paralyzing. The Center for a New American Dream tries to instill this idea through two programs that raise awareness of unsustainable consumption and prompt individual action: Turn the Tide and Step by Step. Individuals who join Turn the Tide commit to taking nine simple actions that protect the environment, such as reducing the carbon we put in the atmosphere by skipping trips in the car. The center's Step by Step program enables people to write and sign onto letters that encourage businesses, communities, and government to change the way goods are produced and consumed. By creating these and similar campaigns, **community organizers** are mobilizing citizens to join together to change specific policies or corporate practices.

Whether using old-fashioned organizing or media campaigns, public education efforts must involve children, because it will take 50 to 100 years to create a much more sustainable society. Today, kids are hyper-consumers who are often targeted by age two to form brand-name loyalty. Educating and organizing children to be actors in this dialogue will be a key part of changing our society's unsustainable consumption habits. For this reason, there is, and will continue to be, a large need for educators and teachers to bring these messages into schools.

• Galvanizing Business Leaders

Ultimately, we must change production practices as well as consumer practices. To do this, we'll need to set limits on the marketing and advertising that drives rising consumption. There are numerous efforts under way to galvanize business for this cause. Some groups are working cooperatively with business to help them redesign goods to minimize their environmental impact—a literal reinvention of manufactured products is necessary to transform our unsustainable society. Some groups are working in more adversarial ways to pressure companies to change production processes to stop producing goods that hurt people and the earth. Campaigns focused on companies like McDonald's, Home Depot, Staples, and Ford are good examples. Others are organizing against major commercial forces. The legislation creating a "do not call" registry, established to block home phone solicitations by marketers, is a terrific example of how **lawyers** and **lobbyists** can help to change the playing field.

Still others are working cooperatively with enlightened marketing and public relations firms to help conduct social marketing campaigns to educate millions of Americans about the importance of living and consuming consciously, ultimately aiming to create citizen-demanded changes. For example, consumers played a large part in building the organic food industry; cutting-edge businesses responded to growing demands for healthy and safe food, and now the industry is growing at a rate of 20 percent a year, making it the fastest growing part of the food sector. Similarly, the anti-sweatshop movement, primarily led by student activists, has created corporate accountability and social responsibility in the apparel industry.

Targeting businesses in these ways allows us to use the power of the dominant market system to change the ways businesses operate and also increase demand for sustainably produced products. Business, lobbying, marketing, and organizing professionals will play an important role in these efforts.

• Reforming Businesses

In addition to targeting business leaders, we should also work with them to determine long-term processes for sourcing materials, subcontracting suppliers, and measuring profits. Businesses themselves are changing. In doing so, we're beginning to imagine a different world—one where innovation and procurement can help decrease the impacts of unsustainable consumption on the natural world.

Technological Innovation

Technological innovations show the most potential for significantly changing business practices. As mentioned before, some companies are radically changing production processes by reinventing products that are completely recyclable, where all inputs are made from natural, nontoxic, biodegradable, and reused materials. Additionally, product engineers are designing products and production lines where synthetic or chemical waste will never actually be exposed to the natural world because it will always be reused. This life cycle approach can be seen in completely biodegradable and compostable carpets and fabrics that are already on the market.

GREEN MARKETER

AT A GLANCE

Employment:

7,500

Demand:

Rising rapidly

Breakdown:

Private sector, 100 percent

Trends:

- At most smaller firms, marketing duties are distributed among many people. Only large, leading "green" companies have separate marketing departments and staff.
- Green marketing campaigns are often linked with environmental and conservation organizations and goals, making job opportunities in this field diverse.

Salary:

Entry-level salaries start around $25,000, with median annual incomes reaching around $50,000, but top earners can exceed $100,000.

JOB DESCRIPTION

Eye-catching packaging and effective advertising make products sell. Green marketers look for innovative ways to introduce environmentally conscious products, such as cleaning supplies, organic food, and natural care products, into the market.

While green marketers need not be artistic geniuses, they must have an eye for design schemes that make products sell. They help plan, design, and implement print, television, and website advertising with other marketing team members. They also contract with graphic designers, who transfer creative visions to realistic television and print ads.

Green marketers are constantly looking for what motivates people, so they often research consumer demand. Janice Shade, a green marketer at Seventh Generation, says that she constantly analyzes sales data, including consumer spending habits and distribution statistics, to evaluate the company's advertising efforts and incorporate successes into their overall business plan. She also writes reports that are presented to the company's executive management team and finance department.

With this research, green marketers decide where to advertise and who their target audience will be. Ideally, they hope that everyone will jump on the "green" bandwagon, yet they also realize that not everyone adopts environmentally conscious consumption habits. "We target our product line to people who don't go out of their way to buy environmentally sound products but still realize that their personal actions are related to the environment," Shade says. Green marketers cater to these people by looking for ways to introduce products into mainstream stores, instead of only specialty shops and catalogs.

GETTING STARTED

• Employers look for job applicants with a bachelor's degree in marketing or business, although psychology and sociology are also acceptable courses of study.

• An M.B.A. is helpful but not required for newcomers to the field.

• Developing written material, helping with fundraisers, and learning "people skills" are valuable experiences for people interested in marketing.

ADVICE FROM THE PROS

Brandi West, website and production manager for Simple Green in California, says she tries to sell her company's nontoxic cleaner to consumers who are more apt to buy toxic brands like Clorox or Pine-Sol. One strategy is to target parents in order to convince them to purchase nontoxic products as another means of baby-proofing houses. "Selling an environmental-friendly product is not enough to convince consumers. You have to show an added benefit. To do this, you need a great business sense and a passion for the product you're selling. If you don't have that passion, it will always be an uphill battle."

West, who earned a psychology degree in college, also advises working in advertising or marketing research for a year or two before attempting to secure a green marketer position. "This way you learn about competition, how companies market their products and how the consumer market operates."

Rapid technological innovation like this will lead to many positive changes in our lifetime. We will transcend the combustion engine and disconnect from fossil fuel. We will be able to grow and distribute more food locally. We will use recycled products to build, mend, and protect our local infrastructures. This revolution in production will generate new jobs for all types of professionals, including industrial designers, materials scientists, and engineers who will rethink everything from how to redesign a car to how to fashion a dress. Economists will measure our national happiness index—not merely our consumer spending. Socially and environmentally responsible companies will win government contracts and market share.

Environmentally Preferable Purchasing

As products are reinvented, individual and institutional consumers can consciously purchase sustainable products. This will have immense benefits to society and the environment. For example, when government agencies redirect millions of dollars to hybrid electric cars for their fleets, we stem global warming and improve air quality. When companies buy 100 percent postconsumer waste paper rather than bleached or virgin paper, they conserve forests, which in turn provides many important ecosystem services like filtering air and water and providing habitat for animals and endangered plants.

Companies often complain that they don't have enough "green" customers. They can't afford to change production and harvesting techniques if nobody buys the new and better products. One key lever for change is building consumer demand for socially and environmentally responsible goods. When big institutional buyers use their purchasing dollars to buy environmentally preferred products, suppliers take notice and begin to respond.

Creating a Cultural Renaissance

Changing policy, mobilizing citizens, targeting business leaders, and reforming business are all strategies to combat our overworked, overstuffed, and overstressed way of life, where society and culture are dominated by commerce, material pursuits, and consumption. We must step back and determine what life is for, what really matters.

Part of the effort to create a safe and beautiful world is reclaiming our culture. Increasingly, people have the urge to pull back from doing more to salvage the nonmaterial, nonwork dimensions of life. We are seeing a cultural and spir-

Career Spotlight

Co-op America
Washington, D.C.

If there is an epicenter for the nation's growing network of green consumers, it may very well be 1612 K Street, in Washington, D.C., the headquarters of Co-op America. With only thirty-five staff people, Co-op America has become an indispensable resource for citizens and business who have made a commitment to rethinking their consumption practices and supporting more environmentally and socially responsible ways of living. The organization also promotes buying from and investing in "green" companies as one way of instigating needed change in the corporate world.

Co-op America's ambitious strategy has earned the group more than 50,000 members and an audience of hundreds of thousands more who visit the group's issue-specific websites or buy its publications. Serving their needs is the job of marketing and consumer programs coordinator Amanda Romero. "Good information is critical for making environmentally beneficial choices," Romero says. "My job is to identify industries that consumers have special interest in and create programs to deliver the facts that people want." With a B.S. degree in resource and environmental studies, Romero began as an intern at the company, as several Co-op America staff have done.

With such a small staff, it's remarkable how many different programs and projects Co-op America is able to sustain. The consumer education staff, for example, publishes the popular *National Green Pages*, also available online, and maintains the online "Responsible Shopper" feature that allows consumers to investigate the records of hundreds of companies on factors like sweatshops, ethics, discrimination, pollution and more.

Other staff members coordinate conferences, political campaigns, and the company's *Co-op America Quarterly* publication. They also oversee the organization's Sustainable Community Program, which combats predatory lending and works to increase community investing.

This Co-op America–sponsored fair and others like it promote fair trade by giving consumers a chance to buy directly from craftspeople in developing countries.

368

The spirit of Co-op America is embodied in its leader—executive director Alisa Gravitz. For fifteen years, Gravitz has helped lead a movement aimed at creating a truly sustainable economy through consumer and citizen action. Besides gathering a strong membership, Gravitz has also organized a consumer-citizen force of more than a million people to urge corporations to improve their business practices.

Among other accomplishments, Gravitz formed the Co-op America Business Network of 2,500 responsible companies and co-authored Co-op America's acclaimed guide to social investing, *Financial Planning Handbook for Socially Responsible Investing*. Gravitz comes to her work at Co-op America with great credentials—a master's degree in business administration from Harvard Business School, with a concentration in marketing and finance.

In addition to her job as executive director, Gravitz is co-founder and vice president of the Social Investment Forum, a national trade association for socially responsible investment professionals. In this position, Gravitz played a key role in the growth of what is now a $2.2 trillion industry.

Other Co-op America employees also share the goal of reshaping the corporate world for the better. Frank Locantore, the company's WoodWise/PAPER project director, runs the Magazine PAPER Project. His main responsibility is convincing the magazine industry to switch to postconsumer recycled paper, or PCR. The first order of business is developing a level of trust from the industry, building support from other nongovernmental organizations, and converting large-circulation magazines to PCR.

"This job requires me to be able to switch back and forth between different skills," says Locantore, "such as negotiating, research, writing, consensus building, grassroots outreach, and advocacy." All of those skills have been called upon in Locantore's formation with other leaders of the Environmental Paper Network. This network "strives to identify and implement a systems-wide approach to transforming the destructive paper production industry towards sustainability."

Deanna Tilden, data entry specialist, has worked at Co-op America for seventeen years. She handles individual and business member data entry, including Social Investment Forum members, which includes new records, renewals, special appeals, sustainers, partners, leaders, and even complaints.

As a deaf person, Tilden says she is pleased with Co-op America's effort to provide sign language interpreters during meetings. She also enjoys the company's sensitivity to its employees' health conditions and the fact that the office has chemical-free carpet cleaning, lights that are energy-efficient, comprehensive recycling, and food waste composting. "At Co-op America, we really walk the talk," she says with pride.

itual renaissance manifested in the popularity of meditation, yoga, book clubs, dance groups, and poetry slams. Increasingly, neighborhoods are organizing, citizens are volunteering, faith-based institutions are tackling local problems, and civic renewal organizations are emerging. These activities are a very healthy reaction to a society and culture that are overly dominated by consumerism and commercial definitions of the good life. Reclaiming our own culture in these ways will allow us to individually and collectively live more fulfilling lives.

Career Advice from the Expert

Recent college graduates looking to enter the sustainable consumption field should become as acquainted as possible with the core concepts discussed in this book. Perhaps the most important qualification for this field is a deep commitment to a different kind of future. Further, the most successful candidates for jobs will probably be experimenting with change in their own lives—examining their own attachments, consumer choices, and inner yearnings.

This field needs people with a variety of skills and backgrounds, as there is no particular academic preparation necessary for obtaining an entry-level job. As with nearly all social change jobs, it's helpful to have strong communication skills—both oral and written. Those with experience or knowledge of politics, organizing, green purchasing, product development, advertising, marketing, or teaching may have advantages because they already have firsthand knowledge of the difficulties this field can present in accomplishing change. Those with backgrounds in communications, legislation, or economics will also have an advantage because they will have worked in areas that are directly applicable for accomplishing change.

That being said, prospective professionals will benefit from internship or volunteer experiences. You should not underestimate these experiences because they may lead to a career. At the Center for a New American Dream, we have hired interns in the past as full-time employees after their internships ended because they were fortunate enough to be involved in our work when staffing needs changed. While this is not always the case at the center or any other organization, internship or volunteer experience will at least demonstrate your interest and commitment to working in the sustainable consumption field.

Overall, there are numerous opportunities for young adults interested in this work. The center looks to hire people who have a love of invention, a willingness to ask hard questions, and a demonstrated capacity to work well with groups on problem solving. Remember, nobody knows how to create a sustainable society; we're inventing it as we go. This requires people with patience, a sense of humor, and a hunger for change.

RESOURCES

Center for a New American Dream, www.newdream.org

Co-op America, www.coopamerica.org

Greenbiz, www.greenbiz.com

Green Seal, www.greenseal.org

Move-On, www.moveon.org

Responsible Shopper, www.responsibleshopper.org

Rocky Mountain Institute, www.rmi.org

Socially Responsible Investment Forum, www.socialinvest.org

The Natural Step, www.naturalstep.org

World Resources Institute, www.wri.org

19. TOXICS AND THE ENVIRONMENT

Dr. Theo Colborn.

A CONVERSATION WITH
Theo Colborn

After gathering and systematically analyzing health-related literature from wildlife, laboratory, and human research, Dr. Theo Colborn has a well-earned international reputation for proving that chemicals pose hazards to humans and wildlife.

As one of the world's leading authorities on chemicals that interfere with the endocrine system, which controls the body's metabolic activity, Colborn has demonstrated that traditional toxicological testing does not detect many endocrine disruptors and that widespread dysfunction is already affecting some populations. "When facing these realities, most scientific experts say, 'Why didn't we think of this before?'" she quips. "And that's the state we're in now. If we had, these chemicals would never have been released into the environment."

Colborn holds a B.S. in pharmacy from Rutgers University and a Ph.D. in zoology from the University of Wisconsin. She previously worked in the Office of Technology Assessment on a Congressional Fellowship and is now a professor at the University of Florida and a senior fellow of the World Wildlife Fund. She has published widely and is best known as co-author of *Our Stolen Future* (Plume, 1996).

What Is the Issue?

A vast number of widely used industrial and agricultural chemicals that were once considered safe are penetrating the womb (and the egg in birds, reptiles, and fish) and impairing the development of embryos and fetuses. These chemicals—called environmental hormones or endocrine disruptors —interfere with the production, transport, signaling, metabolism, and excretion of hormones that are critical for normal development from conception to birth, and the regulation of body function throughout life.

- Adverse Developmental Effects

Endocrine disruptors are not immediately lethal. Their damage is usually not reversible and not always visible at birth—the disorders they cause may not become evident until adulthood. As more and more synthetic chemicals have come into use, so has the number of human disorders increased. For example, human epidemiological studies since the 1970s show that the U.S. population is experiencing increases in the prevalence of childhood disorders such as autism, attention deficit hyperactivity disorder, cancers, diabetes, arthritis, autoimmune problems, hypospadias, early puberty, and juvenile delinquency. As the population ages, it is experiencing increases in early testicular cancers, obesity, adult diabetes, endometriosis, benign prostate hyperplasia, male and female sex organ cancers, reduced fertility, and Alzheimer's disease. Most of these problems are endocrine-related and many have been traced back to fetal origin. Biological changes and symptoms associated with many of the disorders have been induced in laboratory animals during development with synthetic chemicals.

Many have argued that although many of these chemicals might have adverse health effects, the concentrations at which people are exposed are so low that they are not harmful. However, it's been demonstrated repeatedly in the last decade that natural hormones (controlling development and function from conception and throughout life) operate at concentrations in the range of parts per billion (ppb) or parts per trillion (ppt) or less, which is much lower than the concentrations of some synthetic chemicals found in humans and wildlife ppb or parts per million (ppm). Unfortunately, the list of known endocrine disruptors is growing. This list includes industrial chemicals that have become an integral part of our lifestyle and economy, such as certain plastic components, fire retardants, new construction materials, cosmetics, perfumes, and some pesticides.

There is hardly anything we use or touch today that does not contain a synthetic chemical. Almost every synthetic endocrine disruptor was derived from by-products of refining crude oil to make gasoline. Because these by-products are so ubiquitous, it's no wonder that they have entered our bodies. They are often in the air we breathe. Some of these chemicals are persistent and build up in body tissues. Others may not be so persistent but are so widely used in everyday products that exposure is continuous. And in every case, the exposure mixtures change from day to day. There are over 87,000 chemicals in use today that have never been tested for endocrine effects. If only one chemical were involved, reducing exposure would be relatively simple. However, this is not the case, and solving this problem requires a global effort, because of international trade and commerce.

- ## Inadequate Testing

Standard toxicological testing is not sensitive enough to detect the insidious, chronic, and systemic effects of endocrine disruptors that are increasingly reported in humans and laboratory animals. Currently, if a chemical is tested at all, it's done at high doses for obvious problems like birth defects, mutations, convulsions, skin and eye irritation, and cancer. From those high-dose results, scientists have extrapolated down to a level of exposure that is considered safe. Unfortunately, the damage a chemical inflicts in a fully mature animal does not predict the damage in a developing animal before birth or shortly thereafter. Consequently, current assays are not detecting the constriction flaws in our children that can lead to the growing list of disorders reported around the world. The government mainly relies on chemical producers to provide data determining safe exposure levels for specific chemicals. To date, few chemicals have been tested for their endocrine system-disrupting effects because standard protocols have not yet been validated.

- ## How Chemicals Enter Living Beings

The release of chlorinated chemicals such as polychlorinated biphenyls (PCBs) into aquatic systems since the mid-1920s provides well-documented evidence of how toxic chemicals become dispersed in the environment and ultimately end up in our bodies. PCBs in particular love fat and do not like water. When their molecules enter aquatic systems, they latch onto fat in microscopic organisms like algae, and a single microscopic filter feeder can pick up as many as 400 PCB molecules in one day just from ingesting algae alone. These filter feeders are then swallowed by larger filter feeders, and so on, up to the largest fish in the aquatic system, which humans or birds often eat. Each successive animal tier in the food web therefore carries more and more PCBs, so that eventually animals at the top of the food web hold as much as 100 million times more PCBs than the original lake water contained.

This phenomenon is known as biomagnification, because the amount of chemicals has been magnified or exponentially increased in animals at the top of the food chain compared to the original chemical exposure in the natural environment—in this case the lake water. As biomagnification became better understood among wildlife biologists, they realized that, although adult animals appeared to be relatively unharmed by PCBs in their bodies, their offspring were not surviving. Often, if they did survive, they had numerous developmental problems. The seas and oceans were also becoming contaminated, as demonstrated by the fact that whales, more than any other species, accumulated the highest concentrations of some chemicals in their fat. As mother whales nursed their young, they unloaded fat-loving chemicals into their babies through breast milk. It turns out that gestation and breast-feeding are responsible for passing PCBs and many other chemicals from one generation to the next, in humans as well.

- Persistence

About fifty years ago, **chemists** were beginning to detect some of these persistent organochlorine chemicals in wildlife tissue because concentrations were so high. Soon after, they discovered such chemicals in human tissue as well. Yet, here we are fifty years later, still facing the same problem on a global scale. PCBs are still everywhere **chemists** look. And PCBs are only one of the persistent chemicals that threaten future generations.

How Are Environmental Professionals Approaching the Issue?

The concept of endocrine disruption was introduced in the book *Chemically-Induced Alterations in Sexual and Functional Development: The Wildlife/Human Connection* (Princeton Scientific Publishing, 1992). The idea immediately caught the interest of academics around the world, from those working at molecular, cellular, tissue, and organismal levels, to both basic and applied scientists. The greatest enlightenment came when basic scientists working with

BOX 19-1

ADDRESSING TOXICS

- Developing More Effective Testing
- Identifying At-Risk Populations
- Putting Restrictions on Persistent Chemicals
- Creating a Research Agenda
- Involving Business and Industry
- Continuing the Search for More Information

natural systems or organs began collaborating with **chemists**, physiologists, statisticians, and medical doctors as they introduced endocrine disruptors into the systems they were studying.

- Developing More Effective Testing

As this approach matured, it became apparent that traditional high-dose testing yielded misleading results and that testing had to be conducted at the concentrations that synthetic chemicals appear in the environment, in animal and human tissue, and during development. As research becomes more refined and merges with new high-tech instrumentation that can actually produce pictures showing the earliest stages of cellular, tissue, and organ development, the endocrine system's extreme complexity has become more and more evident. We are now only in the earliest stages of understanding the life-giving processes that assure the integrity and survival of animals on the earth, including humans.

For example, let's look at what has taken place within only the past year in several laboratories investigating the effects of the synthetic chemical, bisphenol A. Bisphenol-A (BPA) is a simple molecular unit used to manufacture polycarbonates and similar plastic resins. Every compact disk is made with BPA. It's also used to line food cans; to make baby bottles, high-impact glass, and sporting equipment; and as a fire retardant in many household products.

When BPA, a weak estrogen, is given to a pregnant rodent, it enters developing cells in her fetuses' mammary and prostate glands and causes these cells to proliferate as if they were exposed to more powerful estrogens like estradiol or diethylstilbestrol (DES). Much to everyone's surprise, BPA turns on auxiliary enzyme systems in the cells, or in the interior surface of cells, and catalyzes the proliferate response. BPA also causes these glands to develop unusual ductal systems that are embedded with excessive numbers of estrogen receptors. These additional receptors sensitize the glands to estrogen-like compounds later in life and cause cancer. Had we depended upon simple test-tube tests to examine BPA, we never would have learned about its potential adverse affects. In simple cell tests, BPA simply behaved like a very weak estrogen, which did not predict this long-term, delayed effect that was initiated while the individual was in the womb.

As interest in BPA has soared, environmental **chemists** have shown that BPA is so widely used in common household products that daily low-dose exposure is the norm. Leaching of BPA from plastic containers, baby bottles, and BPA-lined cans has been demonstrated. Independent physician teams throughout the northern hemisphere have found BPA in amniotic fluid consistently in the 2 to 3 ppb range, and in fetuses at higher concentrations. BPA has already been shown to affect the earliest developmental stages of critical areas of the brain that involve information processing and the ability to relate to others and behave as a compassionate human being.

Additionally, more and more and laboratory studies have pinpointed the specific days or stages throughout gestation when endocrine-disrupting chemicals can cause damage leading to developmental and functional impairment in offspring later in life. This type of evidence is helping shape new protocols for testing. It also reveals the urgency for developing new testing protocols as soon as possible, especially since chemical production and use are rapidly increasing around the world.

Although much publicity at first was given to sex hormones, exploration has broadened extensively into the very complex and resilient thyroid system. And more endocrine disruption literature is appearing with regard to damage to immune, pancreatic, and adrenal systems. Scientists are also focusing intensely on chemicals that could cause obesity by interfering interfere with glucose metabolism, insulin, and body weight—end points, which when testing for safety, were generally ignored in the past.

ENVIRONMENTAL HEALTH SPECIALIST

AT A GLANCE

Employment:

68,000

Demand:

Excellent

Breakdown:

Public sector, 70 percent

Private sector, 30 percent

Trends

• Increased focus on environmental justice issues will increase demand for environmental health specialists with a strong commitment to promoting healthy communities.

• Widespread emerging pathogens, such as the West Nile virus, are increasing the need for professionals with strong backgrounds in epidemiology.

Salary:

Entry-level salaries begin at $30,000, with median annual earnings at $47,500. Most earn between $36,500 and $62,500, with the top 10 percent earning $78,500.

JOB DESCRIPTION

Environmental health specialists work to minimize the impact of environmental factors on the health of the public and the natural environment. They assess conditions of food service establishment sanitation; storage and disposal of toxic and hazardous waste; water, air, and land pollution; and the presence of animals and insects. "It's fascinating work on a day-to-day basis," says Brian Collins, director of environmental health for the city of Plano, Texas. "You're overseeing consumer health and safety, and making sure chemistry is safe for the environment."

Local and state government public health departments often employ workers known as sanitarians, environmental health practitioners, compliance officers, and health inspectors. These professionals monitor compliance with health codes and environmental laws, enforcing laws such as smoke-free ordinances and proper waste disposal. For example, environmental health specialists must communicate effectively with restaurant owners, supervisors of sewage and water treatment facilities, and foremen at construction and building sites. "The idea is to help people understand the health implications of what they're doing and get them to understand why they need to comply with health codes and laws," says Randy Grove, environmental sanitarian for Anchorage's Food Safety and Sanitation Department in Alaska.

At times, environmental health specialists implement community education programs or write reports documenting observations, which sometimes results in corrective action or revised local or state ordinances. While they mostly work in local and state public health departments, some professionals find employment in universities, hospitals, and private companies.

GETTING STARTED

• A bachelor's degree in environmental health or a related science such as toxicology, epidemiology, or industrial hygiene is required for employment.
• A master's degree or Ph.D. is helpful, especially for professionals interested in research.
• Most states require health specialists to be registered. Contact your local health department for details.
• Accreditation by the National Environmental Health Association (www.neha.org) is highly regarded.
• Work in local health departments or government agencies like the Center for Disease Control provides a good introduction to the workings of public environmental health.

ADVICE FROM THE PROS

Dick Pantages, former assistant director of Alameda County's Public Health Department in Oakland, California, says the most important thing he learned on the job were people skills. "We go places where nobody wants to see us, so it's a constant battle to get people to comply and understand health laws," he says. "We must explain that following the law is better for their business. The main thing to remember is that there are many ways of approaching people."

- Identifying At-Risk Populations

Epidemiologists and public health authorities are attempting to identify at-risk populations for endocrine disruptors in order to determine how to avoid exposure. At the current U.S. birth rate, at any one time about 1.1 percent of the population is at risk for endocrine disruption—the cohort of unborn children in their mothers' wombs. In addition, newborn children are also at high risk because their development is incomplete and they spend a great deal of time confined indoors, where many toxic chemicals are now part of our households, often found at concentrations 10- to 100-fold higher than they are outdoors. Some of these chemicals were removed from the market years ago, but they do not break down readily in indoor environments. As mentioned earlier, some are fire retardants required in many construction materials and household products, including flame-proofed clothing and bedding.

- Putting Restrictions on Persistent Chemicals

It's important to note that governments are beginning to consider placing restrictions on persistent chemicals. Certainly, society does not want to mistakenly allow more persistent chemicals to become a permanent part of our biosphere. Unfortunately, PCBs and other chlorinated, brominated, and fluorinated compounds like those found in Teflon, commercial juices, processed cereals, beer, wine, tea, seafood, toothpaste, and drinking water are proving to be far more persistent than anyone realized. Disturbingly, it's now recognized that fluorinated compounds that have served as efficient water- and soil-resistant products in the past have a geological lifespan. They will be around as long as rocks. None has a clean bill of health. Teams of environmental **chemists** are still improving technology to detect these fluorinated compounds in environmental samples and wildlife and human tissue. However, they have already found fluorine-containing compounds wherever they look in animals and humans.

- Creating a Research Agenda

It's not unusual for scientists to devote years doing detective work to get to the root of a problem. During this time, they often improve protocols by taking advantage of new technology that provides more sensitive tools for chemical detection and more sensitive health end points.

For example, a large interdisciplinary team of scientists recently demonstrated that dioxin levels in the water in the Great Lakes in the 1930s and

1940s were so high that lake trout and herring could not reproduce there anymore. Part of the team drilled cores in the lakes to measure dioxin in sediments and from that determined the amount of dioxin in the water in the past. Others worked in laboratories to determine if there was any level of dioxin in the water at which fish could survive and reproduce. Overall, the team demonstrated that the concentrations of dioxin in the lakes would not enable offspring to survive through the "fry stage," effectively explaining why trout and herring populations crashed in the Great Lakes during the 1930s and 1940s. Up until last year, overfishing, predation by lamprey eels, and habitat destruction were the only explanations for the loss of economically important fisheries in this area. Now we know more.

Creating a research agenda to better understand how toxics affect human health is critical. Institutions like the National Science Foundation, the National Institutes of Health, and the Environmental Protection Agency should take the lead in this effort. This research must be a collaborative effort where government and industry are heavily involved in ways that have the public's best interest in mind.

• Involving Business and Industry

For industry as a whole, it would not cost very much to support an independently designed and managed research program. But while it appears that corporations are willing to participate in discussions about removing endocrine disruptors from the environment, how far they will go to cooperate remains to be seen. Fortunately, some corporations are moving ahead to develop their own independent screens and assays to test products. For example, pharmaceutical companies have screened their products for many years to assess side effects (including endocrine disruption) before putting them on the market. However, most industries developing their own testing mechanisms are not sharing protocols. The pharmaceutical industry is no exception; it has never been willing to share its screens and assays with health agencies. Such privacy severely obstructs our overall ability to test chemicals and products that are potentially harmful to humans and ecosystems. For this reason, government should become a more active participant in these processes.

• Continuing the Search for More Information

Regardless of other stakeholders' lack of involvement or collaboration, academic scientists are working to determine the truth about toxic chemicals.

CHEMIST

AT A GLANCE

Employment:

89,000

Demand:

Strong

Breakdown:

Public sector, 25 percent

Private sector, 65 percent

Nonprofits, 10 percent

Trends:

• Increased government regulation is forcing companies to emphasize compliance and environmental processes, creating more job opportunities.

• Opportunities are opening up to work outside the traditional lab setting in fields such as law, health and safety, business, regulation, and public policy.

Salary:

Entry-level salaries begin at $35,000, with annual median earnings at $55,000. Most chemists earn between $40,000 and $70,000, with the top 10 percent earning more than $90,000.

JOB DESCRIPTION

The work of environmental chemists is vital to the health of the environment. According to the American Chemical Society (www.chemistry.org), the fate of chemicals in the environment and their effects are a matter of increasing concern to specialists in environmental management. As such, environmental chemists work to develop products and processes that comply with local, state, and federal safety regulations.

Some chemists study the effects of air, water, or soil pollution, while others, such as materials scientists, create new, environmentally sound materials that reduce the need for using virgin sources. These new materials ultimately leave no waste behind after their useful life. Still other chemists are responsible for developing production processes that deliver such environmentally sound products.

Toshi Sugama, a chemist for Brookhaven National Laboratory in Upton, New York, specializes in developing coatings that protect metals from corrosion. Most recently, he developed a cerium coating to prevent chrome from deteriorating and oxidizing into the air. "Curiosity is a must for this field," he says. "You need that curiosity, an interest in how things work, because it takes a lot of study to be a chemist."

Chemists can also be involved in remediation work, where they collect and analyze samples from contaminated sites to determine levels of pollution.

When working in laboratories, chemists use sophisticated equipment including mass spectrometers, chromatographs, and high-powered microscopes to analyze the chemical properties of samples. They also use various computer programs to analyze data and test chemical compounds.

Chemists who are familiar with laws such as the Resource Conservation and Recovery Act and the Clean Air Act work closely with regulators from state and federal environmental protection offices and environmental health and safety managers to develop remediation plans that ensure industry compliance. They can also work for government agencies themselves, as well as in private consulting firms and academia.

GETTING STARTED

• A bachelor's degree in chemistry is the minimum requirement for employment. Coursework in environmental studies, advanced mathematics, and engineering is also relevant.

• Advanced research positions require a master's degree or Ph.D. Graduate and undergraduate programs accredited by the American Chemical Society (www.chemistry.org) are highly regarded. Some schools offer environmental chemistry programs.

• Laboratory experience and fieldwork are invaluable.

ADVICE FROM THE PROS

Dr. George Cobb, environmental chemist and associate professor at the Institute of Environmental and Human Health at Texas Technology University, says he often works with other scientists to complete projects on toxicology and pesticide use. "There are very few projects that I just work on by myself," says Cobb. "You must be able to do teamwork in this field and be open to the advice of scientists in other disciplines."

Scientists monitoring contaminant levels in marine animals have revealed that the oceans, large seas, and gulfs have become the final resting places for many persistent organochlorine chemicals like dioxins and DDT. By carefully measuring the various classes of chemicals in wildlife, it's become possible to trace pollution sources. Information such as this has led to the formation of international accords to rule out the production of certain persistent chemicals.

For example, albatrosses that nest on Midway Island are carrying unexpectedly high levels of dioxins and beginning to show chick impairment. Documentation has shown that the levels of liver enzyme toxicity that were first reported about fifteen years ago in troubled Great Lakes waterbirds still persist today and are now reaching the same levels in albatrosses in the Pacific. The birds pick up partially incinerated floating debris on the ocean surface and return to Midway, where they feed their hungry chicks. Like so many persistent endocrine disruptors, dioxin is carried on the wind and ocean currents that traverse the world. It appears that excessive waste incineration on the Asian continent is contributing to the concentration of dioxins showing up in albatrosses that feed only on the surface of the North Pacific Ocean.

Career Advice from the Expert

Anyone interested in entering the new field of researching and fighting against endocrine disruption should have no trouble finding a niche that will lead to an exciting and fulfilling career. It's imperative that newcomers have strong backgrounds in basic genetics, evolution, organic chemistry, biochemistry, endocrinology, and developmental biology, which may pose a problem because there are not many developmental biologists who have devoted their careers to teaching.

However, there's no need to abandon an already established career. I was a grandmother with a background in pharmacy when I returned to school and obtained my Ph.D. at age fifty-eight. I used my knowledge of pharmacy, zoology, epidemiology, toxicology, and water chemistry to study water pollution in North America's Great Lakes area, which, as mentioned, was attracting attention because of declining wildlife populations and the discovery of developmental, reproductive, behavioral, and immunological abnormalities and deformities in the animals. My research led me to conclude that synthetic chemicals released into the natural environment were concentrating in the food web and, ultimately, disrupting endocrine systems. One day I asked myself, "If this is

Career Spotlight

Toxics Use Reduction Institute
University of Massachusetts at Lowell
Lowell, Massachusetts

Toxics use reduction (TUR) is a proven method for achieving both a cleaner environment and a healthier economy quickly, effectively, and relatively inexpensively. As awareness of the power of TUR grew in the late 1980s, state governments pushed legislation to promote the concept. One of the best efforts was the Toxic Use Reduction Act of 1989, under which the commonwealth of Massachusetts established a requirement and a program to help industries, institutions, and communities to reduce their use of toxic materials. This innovative law created the Toxics Use Reduction Institute (TURI).

Located at the University of Massachusetts at Lowell, TURI employs twenty people and conducts advanced research and education for businesses, institutions, communities, and government agencies. Over the past decade, TURI, the Department of Environmental Protection, and the Office of Technical Assistance have helped Massachusetts industries reduce toxic chemicals used in manufacturing processes by 41 percent, while simultaneously improving competitiveness.

The TURI staff consults with government offi-cials, industry managers, private consultants, public interest groups, community groups, and university faculty to provide specialized research, training, and technical support. Current offerings include training for toxics use reduction planners, peer-to-peer mentoring, and technical research grants.

Research associate Heather Tenney works with TURI's core staff of analysts, engineers, and educators on multidisciplinary research. "My undergraduate work and early experiences in industry led me to question the effects of industrial chemicals on human health and the environment," she recalls.

After receiving a B.S. in manufacturing engineering from Boston University, Tenney earned a master's degree in environmental engineering from Tufts University. "My first course at Tufts—Pollution Prevention Management—showed me the huge potential for a manufacturing professional to make a difference in the environment." As an intern in the Office of Technical Assistance, Tenney conducted site visits to help facilities develop TUR options. Afterwards, she was hired by TURI to measure progress for the program and manage the science advisory board.

Although TURI's professionals work hard to encourage voluntary reduction, it helps that Massachusetts pushes the concept as a central part of its environmental protection strategy. Under state law, firms are required to prepare a plan with features of an environmental management system (EMS), which emphasizes pollution prevention and details management methods and emergency response preparations.

To help planners, Librarian Mary Vidal manages TURI's Technology, Environment and Health Library—one of the most comprehensive pollution prevention libraries in the country, with over 15,000 books, reports, and journal articles, and more than 100 journal and newsletter subscriptions. A searchable "green list" online catalog features publications on toxic chemicals, TUR, pollution prevention, and related information. Vidal, previously a librarian at Air & Water Technologies, received her M.L.S. from Rutgers University and a B.A. in history from Rutgers College.

TURI's surface solutions lab looks beyond current approaches to discover new reduction methods. Lab director Dr. Carole LeBlanc and her team are currently seeking safer alternatives to hazardous organic and chlorinated solvents. With bachelor's degrees in chemistry and biology from Boston College, LeBlanc was a laboratory manager for Gaulin, Inc., prior to joining TURI. She was the first American woman to complete the sustainable development and management program at Erasmus University in the Netherlands, with leading-edge courses such as cleaner products, clean production, and industrial ecology.

Communication is a vital means for TURI to accomplish its mission. Elaine Hays Keough, communications consultant, explains, "We need to reach people on several different levels —consumers to build awareness of what toxics exist, industry to offer safer alternatives to toxics, and state government to explain the value of our work."

To do this, she writes brochures and newsletters, and advises staff on web content and grant proposals. She speaks with reporters about projects and represents the institute at industry events to attract prospective participants and grant applicants. Her bachelor's degrees in English and psychology are both from the University of New Hampshire.

"What I enjoy most about this work," Keough sums up, "is that we get to learn about emerging new areas in science and industry, and then figure out the best way to explain complex ideas so that people can solve real problems in Massachusetts."

happening to wild animals, what are we doing to our own?" From there, the first conference about endocrine disruptors was organized, involving many different professionals who would not necessarily work together in traditional settings. The Wingspread Consensus Statement and the technical book *Chemically-Induced Alterations in Sexual and Functional Development: The Wildlife/Human Connection* emerged from this conference, but this was just the beginning.

From my own story, you can see that this work takes an open mind, an innate curiosity about life, the ability to thrive on discovery, and finding like-minded individuals in other disciplines with whom you can work. Remember, the endocrine disruption field came from the merging of pharmacology, toxicology, molecular and cellular biology, wildlife biology, developmental biology, the medical sciences, chemistry, statistics, and epidemiology, to mention just a few. In addition, endocrine disruption is a very contentious issue. It poses serious questions about responsibility, ethics, morality, the power of the present judicial system to deal with chemicals of this nature, and the economic feasibility of ever being able to do something about what we have wrought. I can think of no career choice that would not allow a person to work in the field of endocrine disruption. This field is especially a good fit for those who love a battle.

Granted, today one of the most limiting factors for removing endocrine-disrupting chemicals from the biosphere is the complete lack of a set of screens and assays to identify and prove they are harmful. Even if such tools were available, however, we would still lack enough knowledgeable technicians, scientists, and laboratories to move forward. If you are bent on scientific research and love detective work, there will be vast opportunities in this field. However, some of the largest hurdles ahead associated with endocrine disruption will be found in the courts, political arenas, and the people's will to act. In order to achieve success in these arenas, it will take brilliant communicators, writers, artists, **lawyers**, philosophers, theologians, and educators to bring this complex issue into the public mind-set, and especially into the minds of those we elect.

RESOURCES

Center for Bioenvironmental Research at Tulane and Xavier Universities, www.cbr.tulane.edu

Center for Health, Environment and Justice, www.chej.org

Children's Environmental Health Network, www.cehn.org

HealthCare Without Harm, www.noharm.org

National Environmental Trust, www.environet.policy.net

National Institute for Environmental Health Sciences' National Toxicology Program, www.ntp-server/niehs.nih.gov

Our Stolen Future, www.ourstolenfuture.org

Physicians for Social Responsibility, www.psr.org

Toxics Use Reduction Institute, http://www.turi.org

U.S. EPA Office of Prevention, Pesticides, and Toxic Substance, www.epa.gov/internet/oppts

World Wildlife Fund's Global Toxics Initiative, www.wwfus.org/toxics

Index